T0073085

Modern Aspects of RELATIVITY

Modern Aspects of RELATIVITY

Eckehard W Mielke
Universidad Autónoma Metropolitana, Mexico

NEW JERSEY • LONDON • SINGAPORE • BEIJING • SHANGHAI • HONG KONG • TAIPEI • CHENNAI • TOKYO

Published by

World Scientific Publishing Co. Pte. Ltd.
5 Toh Tuck Link, Singapore 596224
USA office: 27 Warren Street, Suite 401-402, Hackensack, NJ 07601
UK office: 57 Shelton Street, Covent Garden, London WC2H 9HE

Library of Congress Cataloging-in-Publication Data
Names: Mielke, Eckehard W., author.
Title: Modern aspects of relativity / Eckehard W. Mielke.
Description: New Jersey : World Scientific, [2022] | Includes bibliographical references and index.
Identifiers: LCCN 2021050557 (print) | LCCN 2021050558 (ebook) |
ISBN 9789811244049 (hardcover) | ISBN 9789811244056 (ebook) |
ISBN 9789811244063 (ebook other)
Subjects: LCSH: Relativity (Physics)
Classification: LCC QC173.55 .M54 2022 (print) | LCC QC173.55 (ebook) |
DDC 530.11--dc23/eng/20211201
LC record available at https://lccn.loc.gov/2021050557
LC ebook record available at https://lccn.loc.gov/2021050558

British Library Cataloguing-in-Publication Data
A catalogue record for this book is available from the British Library.

For any available supplementary material, please visit
https://www.worldscientific.com/worldscibooks/10.1142/12476#t=suppl

Typeset by Stallion Press
Email: enquiries@stallionpress.com

Preface

This book emerged from the course *Introduction to Special Relativity*, which has been imparted several times by the author since 2000 at the Metropolitan Autonomous University (UAM), in the Division of Basic Sciences and Engineering (CBI) of the Iztapalapa campus.

As many textbooks on this topic are relatively old, for a modern course it was necessary to enrich it with some of the most recent experiments and astronomical observations, with which it is possible to more accurately verify Einstein's theory.

Today, relativity has been partially applied in many engineering projects, the most notable being GPS devices integrated into mobile phones, cars, and aircraft navigators. While analyzing the collisions of elementary particles at Fermilab or CERN in Switzerland, physicists are playing "Relativistic Billiards" every day. On the other hand, the animation of relativistic moving objects, based on Relativistic Optics, has yielded benefits in computer programs such as "Ray Tracing", among others.

The focus of this book is to cover the need for an elementary text based on the Invariance of Light propagation and spacetime diagrams. We hope that the book will therefore also be suitable for an online study without a teacher, employing (free) computer algebra. In the first chapters, the mathematics used is relatively simple, such as the Pythagorean theorem and the intuitive vector calculus. Later chapters gradually use four-dimensional vectors or even tensors, which are rather elegant, but are not indispensable to understand the basic principles of Relativity.

The enthusiasm and interactive participation of students was the key to these updates in an accessible and fun way. Two of them, Silvia Cortés López and Daniel Martínez Carbajal, have already graduated in physics and were therefore well qualified to assist in the preparation of the book.

More recent updates of the original Spanish lecture notes, published in June 2015, have been assisted by the students Stephanie Catalina González Migoni, Diana Haideé Ramírez Gaytán, and Guadalupe Sagaon Rojas from the Metropolitan Autonomous University. We appreciate the opportunity to participate in this project, as well as for the support at every moment of the process.

The cover is based on a color drawing of my daughter Miryam Sophie.

Mexico City, March of 2021.

Eckehard W. Mielke
Physics Department
Metropolitan Autonomous University
Iztapalapa, Mexico City, Mexico

Contents

Chapter 1

Introduction: "Dawn of Time"

For the prehistoric man, time must have been a rather mystical succession of days and nights, a warning about the existence of cyclic phenomena... the moon changed its shape, etc. Like many animals, humans have a pre-conception of time due to internal "biological clocks" (Droin *et al.*, 2019).

Plato, the greek philosopher, held the view: "Time is the eternal image, ... an abstract idea like a life reality." Marcus Aurelius, the Roman emperor: "Time is like a river that carries away rapidly everything that was born." Saint Agustin (354–439), a bishop and philosopher: "What is time? If nobody asks me, I know it; if I want to explain it to some one, I do not know it."

Forever, the effort to measure time and create a feasible calendar was one of the biggest worries of humans, a riddle for astronomers, mathematicians, priests, kings, and every one who needs to count the days till the next harvest, calculate when people have to pay taxes, or determine the exact moment of a sacrifice to calm down a choleric God.

1.1 The Concept of Time in Ancient Cultures

In Egypt, 4 000 years before Christ, the 365 days of a solar year were already known. The beginning of the year was determined from the first appearance of Sirius, the brightest star at sunrise. This event coincided with the rise of the Nile river.

A moon–sun calendar was established, consisting of 12 months with 29 and 30 days by the Greeks. Heraclitus claimed that all existence is build from motion flow. "You cannot bath two times in the same river, because the water that flows over you always is new." Time for him was like a river, where everything experiences a change.

In Babylon, 500 years B.C., the duration of the year was 365 days, 6 hours and 15 minutes, and the 60 seconds minute was established by the astronomer Nabourianos. The Babylonians had ingenious methods to calculate this, dividing stake shadows into degrees, minutes, and arcsecond angles, and clepsydras or water clocks.

In ancient Rome, the moon–sun year had 10 months; most of them were dedicated to their gods: January came from Jano, the double-faced God that looked to the past and the future. February originates from Latin februa and it refers to purification festivals celebrated by Roman priests, the objective being the soul purification of women. March was named after Mars, a war god. April is derived from aperture, because it is the season when flowers open. May has to be because of Maia, the goddess of spring. June came from Juno, the god of marriage. September, because it was the 7th (Septem) month of the old calendar. At the same time, October and November were the 8th and 9th months. December, the 10th month, was represented by a slave who carried a lit torch in allusion to the Saturday's festival. In 44 B.C., after Julius Caesar's murder, a month was added in his honor: July, the 7th month when he was born. Augustus, Julius Caesar's successor, added another month: August is nowadays the 8th month; then September is the 9th, October 10th, and so on.

Accurate calendars were developed by Mesoamerican civilizations, as well, and that of the Mayas were the most sophisticated.

Calendar precision let the Mayas organize their routine activities and register the passing of time of political as well as religious events that were considered crucial. The Maya calendar has three ways of counting time: The sacred calendar (Tzolkin or Bucxok, with 260 days), the civil calendar (Haab, 365 days), and the long count.

Tzolkin with 260 days was the most common calendar for the Maya people. They used it for agriculture, their religious ceremonies, and their family customs. Haab calendar refers to the Earth's trajectory around the sun in 365 days. The Mayas divided the 365-days year into 18 "months" named Winal with 20 days each and with extra 5 days named Wayed. Also, they counted days from a specific date determining the beginning of the actual Maya age. According to archeologists, it was 12th of August 3113 B.C. in the "long count".

Aztec natives gave a lot of importance to time and created two calendars: One with 365 days, Xihuilt, that determined their religious ceremonies; it

Figure 1.1: Aztec calender.

was formed of 18 months with 20 days and 5 extra days, too. The other, with 260 days named Tonalpohualli, had a kind of mystic meaning. It was found in huge cut stone, nowadays located in the National Anthropology and History Museum in Mexico City. It consists in the union of a series of 20 glyphs with another of 13 numbers, the combination of both series yielded 260 days.

The Inca culture (Peru and Bolivia) had an amaizing development: The Incas knew the synodic revolution of the planets with admirable accuracy, the kuots in the *Quipus* marked the days on the calendar, which consisted of 365 days for a solar year.

1.2 The Clock Story

The word clock, used for the first time in the 14th century, comes from the Latin "clocca" that means bell. Since prehistory, men have measured time. They erected columns of stone in such a way that when a star coincided with its alignment, it indicated an important time or date. The ancient Egyptian obelisks were pillars whose shadow shifted with the passing of the day and marked the hours between dawn and sunset. Men noticed that the shadow varied according to the sun's position, thus the gnomon was born, embedded into the ground, with grooves indicating the different times of the day. The shadow of the cane indicated different schedules, but

Kind of clock	Year (aprox.)	Developed in/by	Performance	Image
Sundial	3500 B.C.	Egypt	The shadow cast by a gnomon on a surface with a scale is employed during the daytime movement of the sun. The science resposible for gathering knowledge about sundials is called gnomonics.	
Clepsydra	1400 B.C.	Egypt	This clock is named Clepsydra: It is composed of two containers of water: When water flows from the higher container to the lower one, the water level rises up a float attached to a stick with grimaces.	
Hourglass	1338	Marine sand glass	At the beginning, the lower bulb remains static, loaded with sand, while the higher one remains empty. When the clock is turned, the bulb that contains the granular matter is above and the count starts. The sand begins to flow towards the lower bulb because of gravity.	
Pendulum clock	1656	Cristian Huygens	In these clocks, the operation is regulated by a pendulum that oscillates. The driving force is gravity acting on a mass suspended from a rope wound around a cylinder. This is transmitted to a pinion that moves the wheel turning the hands of the clock.	

Kind of clock	Year (aprox.)	Developed in/by	Performance	Image
Marine chronometer	1761	John Harrison	Uses tempered steel for the balance spring and a bimetallic temperature compensation. It allowed navigators to accurately assess the ship's position in longitude.	
Quartz clock	1920	Warren Marrison and J.W. Horton	Quartz is a type of SiO_2 crystal which when subjected to a voltage oscillates at a constant frequency of 32768 Hz (piezoelectric effect). Such a crystal, cut into the shape of a small tuning fork, plays the role of regulator and frequency stabilizer, thereby measuring time. The vibration of the fork produced by the circuit generates an electrical signal of the same frequency or some fraction of it.	
Atomic clock	1949	United States	This type of a clock uses atomic resonance to measure time, feeding its counter based on cycles or radiation frequency. The first devices were masers (from the English "microwave amplification by stimulated emission of radiation"), an artifact that produces electromagnetic waves through stimulated emission. Devices currently built are based on much more advance physical principles related to cold atoms and atomic sources. The precision of the atomic clock is better than one-trillionth of a second and is used as a reference to measure time. The masers of the atomic clock use chambers with ionized gas, usually Caesium, which is officially used in the definition of a second. Since 1967, the International System of Units has defined a second as 9 192 361 770 radiation cycles that correspond to the transition between hyperfine levels of energy of the Caesium-133 isotope.	

they had large inaccuracies. One of the oldest gnomons for which data are available was used in Egypt in 1500 B.C.

In order to take the rotation of the earth and other astronomical observations into account with precision, the solar quadrant was built to improve the precarious gnomon. The former was formed by a disk-shaped stone with marked timelines indicating the different moments of the day. The shadow of the embedded cane on the disk marked the hours. The quadrant was placed in a certain way such that the shadow identifies the same time on any day of the year. The solar quadrant was considered an instrument of greater precision. From this arose the equatorial quadrant and then the universal quadrant, which was accompanied by a compass. Since then it was a useful instrument for navigators. The solar quadrants appeared in Greece around the 5th century B.C. For nocturnal measurements of time, there were stellar and lunar quadrants, but they worked only during a clear and serene sky. Subsequently, new and better instruments for measuring time emerged. The adjacent table summarizes some of the most important clocks:

1.3 The Definition of the Second

In the International System of Units (SI), the unit for the time interval is the second. In 1967, the second was redefined during the General Conference of Weights and Measures as

> *The duration of 9 192 631 770 radiation periods corresponding to the transition between the two hyperfine levels of the basic state of the Caesium atom 133 (nominally at a temperature of 0 K).*

The time interval and its reciprocal, the frequency, can be measured with higher resolution and less uncertainty than any other physical quantity. The definitions of other SI units as the meter, the candle, the Ampere, and the Volt now depend on the definition of the second.

The precise definition was possible thanks to the development of frequency standards based on atomic energy transitions. Since the atomic frequency standards appeared for the first time, after 1950, the science of measurement progressed fast. In 1955, Einstein still showed some interest in the development of atomic clocks (Naumann & Stroke, 1996).

Nowadays, the National Institute of Standards and Technology (NIST) in the USA has the world's most accurate primary frequency standard: A laser-cooled source of Caesium as NIST-F1. Since its completion in 1998,

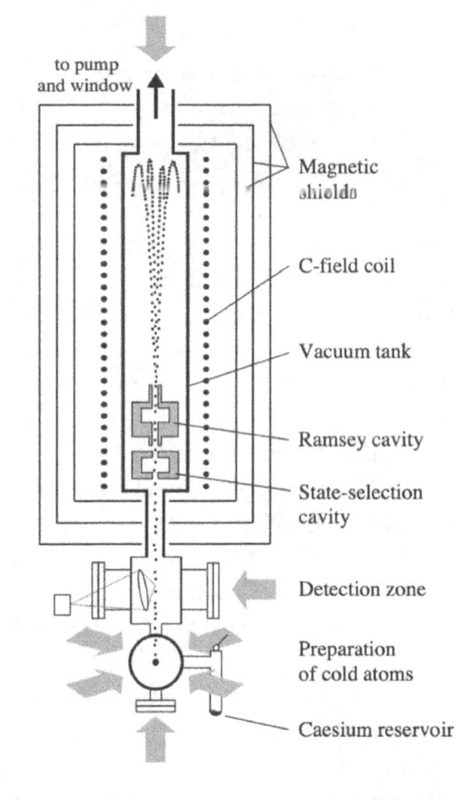

Figure 1.2: Caesium fountain atomic clock NIST-F1.

the NIST-F1 has been continuously improved and now has an accuracy of $\Delta f/f = 4 \times 10^{-16}$. However, NIST-F2 is currently being launched as a second generation atomic standard that promises to be even more precise. Then, it may not "lose a second" in at least 300 million years.

1.3.1 *Precursors of the second*

It is difficult to estimate the impact that atomic measurement of time has on modern society. The technologies that we often use such as smartphones, the Global Positioning System (GPS), and the electric power network depend on the accuracy of atomic clocks.

The redefinition of the second, based on the Caesium atom in the SI, was established in 1967. Earlier, the definition of the second had always been related to astronomical time scales.

1.3.2 *First standards*

All time and frequency standards refer to a recurring event that is repeated almost constantly. Such a periodic event is produced by a device called a resonator, which is driven by an energy source and together, resonator and energy source, form an oscillator. The oscillator works at a certain resonance frequency, which is the inverse of its period T.

All clocks depend on an oscillator, so any uncertainty or change in the frequency of the oscillator will result in a corresponding uncertainty or change in the accuracy of the clock. The performance of an oscillator can be expressed as $\Delta f/f$, where f represents the theoretical value of the frequency and Δf its uncertainty.

Improving the precision with which time is measured has been one of the oldest goals of human cultures. Basically, it is a search for the best oscillators, characterized by well-defined periods, stable and difficult to disturb. At first, astronomers realized that the rotation of the Earth around its axis could serve as a natural oscillator, so the second was defined as a fraction (1/86 400) of the duration of a solar day. Subsequently, to create a more precise time unit, astronomers used the Earth rotation period relative to the Sun to define the second in the SI.

When the second was based on astronomical periods and before the atomic age, mechanical and electrical oscillators served as standards for time measurement in laboratories.

In the National Bureau of Standards (NBS) now known as NIST, pendulum clocks, based on the principle discovered by Galileo, served as the first standards for time measurement. For almost 300 years after Galileo's discovery, pendulum clocks were dominating. Later, quartz oscillators were used, based on the phenomenon of piezoelectricity discovered by Pierre Curie in 1880. They resonated at an almost constant frequency when an electric current was applied. Walter Cady patented a designated piezoelectric resonator as a frequency standard in 1923. These oscillators were soon used to control the radio transmission frequencies.

Quartz oscillators are still in an almost unlimited number of applications. They are present in watches, smartphones, computers, radios, and in many types of electronic circuits. However, their resonance frequency can be affected due to the aging of the crystal and other environmental factors such as humidity or pressure. In addition, the quartz oscillators depend on the crystal shape. This led to the development of atomic clocks. Unlike other oscillators, a group of atomic oscillators theoretically all generate the same frequency. Unlike electrical or mechanical resonators, atoms do not wear out, and their properties do not change with time.

Atomic oscillators use quantized energy levels in atoms and molecules as the source of their resonance. The laws of quantum mechanics and the relativistic Dirac equation (Autschbach, 2014) dictate that the energies of an isolated system, like an atom, have certain discrete values. An electromagnetic field of a particular frequency can drive an atom from one (hyperfine) energy level to another. In the same way, an atom at higher energy can "fall down" to a lower energy level, emitting photons. The resonance frequency of an atomic oscillator is proportional to the difference between both energy levels.

In theory, an atom is a perfect "pendulum" whose oscillations can be used as a frequency standard. Because of that, the current frequency standard NIST-F1, based on oscillations of Caesium atoms, is so accurate. Although it faces two important physical drawbacks: Radiation emitted by the container at 300 K containing the Caesium sample used in the standard and the change in the Caesium frequency (redshift) due to the trajectory that the atoms travel within the cavity, as well as the external gravitational potential. The standard NIST-F2, now under evaluation, seeks to minimize the uncertainty caused by both phenomena.

1.4 What is the Most Accurate Clock in the Universe?

Radioastronomers have discovered more than 17 ms pulsars in our galaxy in a study of high energy sources detected by the Fermi gamma ray telescope.

Although difficult to locate, these types of compact objects like neutron stars (NSs) may serve as a kind of galactic GPS in order to detect gravitational waves that pass close to our planet. A pulsar is a neutron star with a non-aligned magnetic field and fast rotation as a remnant of a massive star explosion. Because the rapid rotation generates emissions of intense gamma rays, radio waves, and particles, these pulsars diminish little by little their rotation as they get older. Old pulsars turn hundreds, almost thousand times per second, even faster than a kitchen blender.[1] These millisecond

[1]Sub-millisecond pulsars of about 1490 Hz rotation frequency may have formed during the coalescence of neutron stars, cf. Yang *et al.* (2019).

pulsars have been transformed and rejuvenated by the increase in matter captured from a companion star.

Radio astronomers discovered the first millisecond pulsar many years ago. To locate these objects with radio traces throughout the sky requires enormous effort and time, and even now only about 300 have been located in the disk of our galaxy.

Millisecond pulsars are the most precise clocks in nature, with a long-term stability below attoseconds, rivaling atomic clocks. The precise monitoring of changes over time in an array of millisecond pulsars throughout the sky (Taylor *et al.*, 2016) may allow to detect nano-Hertz gravitational waves, a prediction of General Relativity.

The Global Positioning System (GPS) is based on the measurement of the arrival time from various satellites to determine your position on

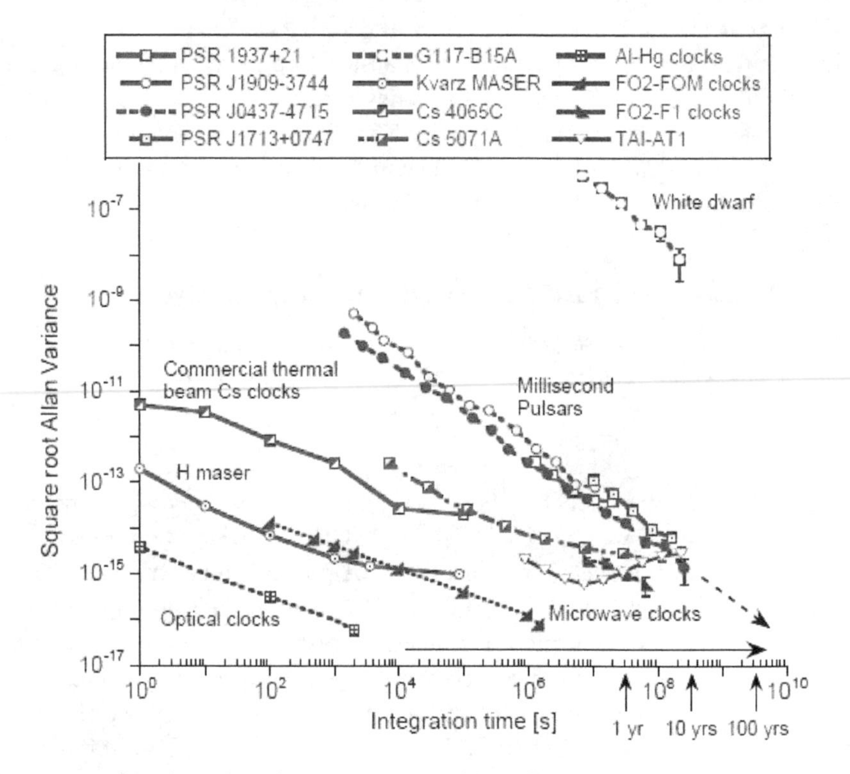

Figure 1.3: Stability of the frequencies emitted by different sources, including those of astrophysical origin (shown in red), frequencies emitted by commercial oscillators (in blue) and frequencies emitted by clocks of the best laboratories in the world (in green), cf. Allan *et al.* (1997).

Earth. Although affected by unpredictable glitches due to re-arrangements of matter inside the NS, monitoring the time changes of a constellation of millisecond pulsars located throughout the sky will allow us to detect the accumulated background of gravitational waves that pass by. For a more detailed view of radio wavelengths, the Fermi Pulse Search Consortium was organized, and a group of radio astronomers with experience in the use of five of the largest radio telescopes in the world: The National Observatory of Radio Radioastronomy; Robert C. Byrd Green Bank Telescope, West Virginia; Parkes Observatory in Australia; the Nançay Radio Telescope in France, the Effelsberg Radio Telescope in Germany, and the collapsed Arecibo Telescope in Puerto Rico. Since 2019, there is a newly operating 500 m telescope in the radioquiet Dawodang depression, China.

After studying approximately 200 objects, and with an intense computational analysis of the data, the discoveries are just beginning. Four of the new objects are pulsars called "black widows", because the radiation comes from recycled pulsars that may have destroyed their companion star to increase their rate of rotation. Some of these stars have their mass reduced to the equivalent of dozens of Jupiters. The known number of these systems in the disk of the galaxy has been increased, which will help us to better understand their formation.

1.5 A Brief Biography of Albert Einstein

Albert Einstein was born on March 14, 1879, in Ulm (Baden-Württemberg, Germany). Pauline Koch and Hermann Einstein were his parents. They married in 1876 and three years later they had their boy-child, Albert. Later, in 1881, his sister Marie was born, whom they affectionately called Maja. Later on, she married a son of the Winters. Albert was fond of their younger daughter.

Hermann Einstein was a cheerful man, fond of life and social environment. Along with his brother Jacob, who was an engineer, they attended a small electromechanical shop. Jacob, being very close to the family, was one of the first persons who influenced his nephew Albert: He transmitted his passion for science and gave him his first math lessons.

Pauline Koch, had a more serious and artistic character than her husband; she was fond of German music in general and Beethoven in particular. Undoubtedly, Pauline constituted a fundamental influence in the development of Albert's personality, since she never doubted the capacity of her son and always had much confidence in him.

Figure 1.4: Albert Einstein, apparently in 1921.

Pauline initiated an interest in music in her son: When he was 6 years old, she hired a female musician to teach him how to play the violin. When Einstein was 13 years old, they started playing together, a custom that they kept up until her death. With time passing on, Albert would become a dedicated lover of Mozart's music. Surprisingly, he was not a prodigy child and, due to his delay in starting to speak, his parents thought he could suffer some kind of mental retardation, which even led them to visit a doctor. This delay has motivated some anecdotes whose veracity has never been clarified: One day during dinner, he unexpectedly started talking and he did it with ease, all he said was that the soup got burned. That he had not spoken until then was because he had nothing to say (?). Even as a child, the most outstanding features of his future personality could be appreciated: Sensitivity, interest in the intellectual, independence, and love of loneliness. He burst into tears at the sight of a military parade and desperately asked to be saved from having — someday — to join the army. On one occasion, while in bed due to a slight illness, his father gave him a compass, an instrument that particularly impressed him. As a student, Einstein did not like to merely memorize, he was of insatiable curiosity, always had questions that were not in the texts, which irritated some of his teachers. He finished primary school in 1888, and a year later he entered the Luitpold Gymnasium, a high school.

Again, economic necessities forced the family to move their business in 1894, this time to Milan. When he arrived there, Einstein continued his studies in the Cantonal school of Aarau, which later allowed him to enter the Zurich Polytechnic School. There Minkowski taught, from whom Einstein learned some geometric tools that were later on essential for his work. Contrary to the German strictness, he liked the Italian educational system. By this time he was an outstanding student in mathematics, although somewhat mediocre in history and classical languages. While still a teenager, he was entertained reading Kant, and already had the firm conviction of becoming a theoretical physicist. As the father's business did not work in this city either, the family had to move to Pavia in the region of Lombardy, Italy.

Devoting himself to physics, instead of mathematics, arose because, by its extension, he found it easier to comprehend the former completely. As a student, he continued to be disliked by the teacher, this time it was Weber who complained: "You are a smart boy, Einstein, very smart, but you have a big flaw: you do not listen to anyone." In 1900, at 21, he finished his higher education and obtained a degree in physics; his average grade was 4.91 points out of 6. The year 1901 would bring many changes for him. He was in charge of the mathematics classes at the Technical School of Winterthur. He published "Consequences of Capillarity Phenomena", his first scientific paper and, with the satisfaction of having found his first job, he wrote a letter to Professor Alfred Stern: What a wonderful sensation one experiences when one discovers the unity of a complex of phenomena which, before sensible perception, appeared as completely independent things! To his luck, he escaped compulsory military service for having flat feet and varicose veins. That year he was also granted Swiss citizenship. The following year, 1902, he worked as a preceptor in a boarding school in Schaffhausen. In June, he began his work as a technician in the Patent Office of Berne, a job he got thanks to the recommendation of the father of Marcel Grossman, a mathematician, co-worker, and friend who contributed his knowledge of Riemannian geometry in the development of General Relativity. According to Albert, the work consisted of doing routine jobs like a "shoemaker"; although later he always recommended young researchers to do this type of work,[2] in order to avoid the dangers of an unbalanced intellect. In 1903, he married Mileva, a student companion, a rather reserved person. Einstein obtained a Doctor of Natural Philosophy degree from the

[2]During 1915 till 1926 he visited Kiel (North Germany) "relatively" often to collaborate with Hermann Anschütz on a gyroscopic compass for navigation, cf. Illy (2012).

University of Zurich for his work "A new determination of the molecular dimensions."

1.5.1 *1905: A special year for physics*

1905 was a special year for Einstein and for physics (for this reason, 2005 had been chosen as the World Year of Physics) because some of his most important works came out then, when he was only 26 years old. In the volume XXVII of the Annalen der Physik, he published "On the electrodynamics of moving bodies." He destroyed the 30 manuscript pages of this article after its publication; but because of historical interest, he made a copy in 1943. In 1906, he wrote an article on the Brownian movement with which he demonstrated the existence of atoms. Fortunately for all, his words: "I will soon reach that stationary and sterile age in which one begins to complain about the revolutionary mentality of young people" were far from being fulfilled, if they ever did. In 1907, he found the famous formula $E_0 = mc^2$ that relates energy, mass, and the velocity of light, and in 1908 published a work about the *Principle of Relativity.* The following year, at 29 years, he was admitted as a professor at the University of Zurich.

In 1910, a vacancy of the University of the Prague occurred and he moved there, this time with the help of Anton Lampa, physicist and enthusiastic disciple of Mach. When Lampa asked Max Planck for references on Einstein's value, Planck responded: "If the theory of Einstein is proven, as I hope, he will be considered as the Copernicus of the twentieth century." That year, his second son, Eduard, was born. Later on, a feeling of abandonment caused his son serious psychological problems. Mileva accompanied their son until his death, after he was internated in a sanatorium. This was one of the biggest family problems Einstein had to endure. Finally, this year Einstein solved the so-called anomaly of the planet Mercury, a problem that had worried scientists for centuries. He published "Influence of the force of gravity on the propagation of light" and, in 1912, "On the thermodynamic bases of the law of photochemical equivalence." He returned to Zurich to occupy the chair of Theoretical Physics of the Polytechnic School, now ETH.

In 1913, he was appointed professor at the University of Berlin and member of the Prussian Academy, as well. Together with Grossman, he published his first work on the General Theory of Relativity. In 1914, against the wishes of Mileva, the family moved to the cold Berlin. Shortly after, the First World War surprised Mileva in one of her summer vacations, who began drifting away from marriage.

During this time, the general situation was unpleasant for Einstein. Obliged during the war to collaborate with the army, like so many other scientists, Einstein participated in an airplane wing design which, to his relief, never got off the ground. At that time, together with Georg Nicolai, he wrote the *Pacific Manifest*, which was signed by a few intellectuals.

A year worthy of being highlighted is 1916, since it was then that he published his famous book on the *Theory of Special and General Relativity.* Later, in 1919, while observing a solar eclipse, the British scientist Arthur Eddington confirmed Einstein's predictions about the curvature of spacetime, and therefore, of light-bending by the sun twice as large as in the Newtonian calculation of Soldner. Then onwards he was idolized by the media.

His outfit was a bit extravagant, and he stopped wearing socks for the rest of his life. The friends of his cousin Elsa, who lived with them, thought he was somewhat crazy, nowadays referred to as the "geek" syndrome, a mild form of Asperger's (Parsons, 2011). Mileva, who knew all this, and since she did not want to share it, made Einstein choose between her or that life, which initiated their divorce. Later on, Einstein married Elsa, in whom he found an ideal partner. In 1921, he was awarded the Nobel Prize in Physics for his theory of the photoelectric effect and not for the theory of relativity.

At that time, in scientific matters, and against the tendency of contemporary scientists, who devote their efforts to quantum physics and statistical mechanics, Einstein continued trying to unify the gravitational and electromagnetic forces.[3] In 1929, the Prussian Academy published his *Unified Field Theory.* In 1933, Einstein embarked on an exile from which he would never return and set his residence in Princeton, because he saw in America a promised land. In 1934, Elsa's daughter Ilse died and, in 1936, his second wife. Together with Leopold Infeld, one of the professors of the late J. Plebansky (Cinvestav, Mexico City), he published the popular book *The Evolution of Physics.*

He advocated the creation of a super-world state with a dissuasive military force. The physicist's support for the creation of the atomic bomb was limited to the writing, in 1939 and in 1940, of two letters to the American

[3] "When Einstein and others tried to unify gravitation with electrodynamics, both theories were classical approximations. In others words, they were wrong. Neither of these theories had the framework of amplitudes that we have found to be so necessary today" (Feynman, 1988).

Figure 1.5: Einstein on a beach of Long Island, 1939.

President Roosevelt to encourage the Manhattan Project. The bombs were dropped from the plane "Enola Gay", when Germans, for economic reasons, had given up their Uranium project (Bethe, 2000). The pilot of the plane finished his days in a psychiatric hospital, and it seems that Einstein, given the devastating results, said that if he had known it, he would have preferred to be a plumber.

In 1950, Princeton University published a new theory of Einstein on the Unified Field: fragmentary, formally coherent, but difficult to corroborate experimentally. In 1952, during another solar eclipse, his Theory of Relativity was confirmed with enough accuracy. This same year they offered him the presidency of the State of Israel, but he presented his resignation arguing that: "Throughout my life, I have dedicated myself to objective problems and lack the natural aptitudes and experience necessary to deal properly with people and exercise official functions."

On April 11, 1955, at the age of 76, Albert Einstein fell seriously ill, he was hospitalized due to an inflammation of his aorta and died on April 18 of the same year. The Funeral was as simple as his tastes: There were no ceremonies, no speeches, not even a grave: Surrounded by a small group of people, he was incinerated and his ashes scattered in the waters of a river.

Few have been the honors that have not surrendered to this genius. He has been considered the most beloved man and the most durable idol on Earth, the Newton and the most powerful brain of his century. Bernard Shaw considered him as one of the eight men who in the last 2500 years have created proper universes of knowledge and represent summits in the intellectual synthesis and discoveries: Pythagoras, Aristotle, Ptolemy, Copernicus, Galileo, Kepler, Newton, and Einstein.

1.6 Einstein and Mileva Maric: A Failed Collaboration?

Mileva Maric was born in 1875 to a mother descendant of Montenegro, in Titrel, a village in Vojvodina, then part of the Austrian–Hungarian empire. As a student at the Royal Gymnasium in Zagreb (Croatia), she obtained permission to study physics in a class for men, graduating with good grades also in mathematics. Then she moved to Zurich and was admitted to the Polytechnic Institute of Zurich, now ETH. Einstein and Maric were the only physics students who entered this section in 1896.

Mileva was Einstein's first wife, the relationship between them progressed during her college years. When Einstein graduated, he got a temporary job outside Zurich, while Mileva continued in college and was preparing for the final exams again which she initially failed. Every Saturday, Albert visited her in Zurich. During one of those visits, Mileva informed him that she was pregnant. The pregnancy affected her studies and she left the university, desolate, returning to her parents' house. Her father, upon learning what had happened, flatly forbade Mileva to marry Einstein. In the winter of 1902, Mileva gave birth to a girl. In letters the couple sent between 1897 and 1903, uncovered in 1987, they referred to her as Lieserl. No one knows what happened to Einstein's only daughter, apparently she developed scarlet fever, but no records have been found. Mileva probably gave her for adoption shortly after her birth, being registered with the name of her new family. In the same year, Einstein moved to Bern to start working in a patent office in Switzerland. Mileva soon followed him without Lieserl and

Figure 1.6: Mileva Maric and A. Einstein.

the couple married on January 1903. About a year later, his first son, Hans Albert, was born. However, Einstein devoted most of his time to work and paid little attention to his wife and son. The situation depressed Mileva, while Einstein took refuge in work.

On July 28, 1910, Eduard, his second son, was born. Things improved among them, but not for long, because Mileva was jealous of the women whom her husband attracted.

In the next years, he was working in the patent office. Especially from 1905 onwards, Einstein produced a constant flow of articles and in 1909 he left the patent office. In October of that same year, marital crisis arose as a result of the jealousy of Mileva toward a friend of Einstein: Anna Meyer-Schmid. The fee that Mileva paid for Einstein's success became evident to those around them: She replaced Einstein's lack of affection by the love for her son Hans Albert, who later became an engineer.

Mileva soon realized that she had to complete not only with science for Einstein: On a visit to Berlin, he started a relation with his cousin Elsa Löwenthal who was divorced and had two daughters. In 1914, Mileva and her children moved to Berlin with Einstein, where she realized that one of Albert's attractions was Elsa, and she returned to Zurich with her two children to never again live with him.

After their separation, Einstein saw Elsa more often and, in September 1917, he went to live with her. Elsa was interested in him and pressured him for divorce. The divorce finally materialized in 1919 in the same year in which they married. Again Einstein's flirtations with other women caused some problems, but the couple remained united until Elsa's death in 1936. By that time they were already living in the United States, as was Maja, Einstein's sister, who upon Elsa's death met in Princeton with Margot Einstein, Elsa's daughter and with Helen Dukas, Einstein's long-term secretary. The three women lived with him, they ran the house and helped Einstein with the correspondence, in addition to offering company, advice, and affection.

Another woman who influenced Albert's life was his contemporary Marie Curie. Her discovery of radioactivity (along with her husband, Pierre, and their colleague, Henri Becquerel) played an important conceptual role in the development of Einstein's famous formula $E_0 = mc^2$. When Einstein and Mileva were still together, they frequented the Curies and became good friends.

About Mileva's work as a physicist, the majority of literature dealt with their stay at the ETH, or referred to her preparation for the second round of examinations. A letter refers to "our work on relative motion." Afterwards, she left Zurich to stay with her parents. In other parts of his letters, he referred to his work in relative movement and other topics of physics, but only to what he had done (Stachel, 1996).

Mileva had undoubtedly begun at the same intellectual level as Einstein; they read, studied, and talked about physics together. However, around 1902, their association may have changed because Einstein's thinking had developed further and was on another level. But until then, it seems that Mileva helped him to concretize ideas.

Initially, the important role that Mileva played was supportive of Einstein, she could have participated by making calculations, looking for information, etc., but she certainly did not contribute to the genius of Albert Einstein, one of the best physicists, who was at the same level as James C. Maxwell or Isaac Newton.

1.6.1 *Ludek Zakel: A son "in theory"*

In 1995, a Czech physicist of then 63 years named Ludek Zakel claimed to be the son of Elsa Löwenthal and Albert Einstein. Although he could not prove it, apparently Albert's stepdaughter confessed to him that his real biological mother was Elsa: She had gone to Prague, believing that she had a tumor, and discovered that she was pregnant. Ludek also stated that he was changed at birth since the real son of the woman, who raised him, died at birth.

Figure 1.7: Ludek Zakel has an amazing resemblance to Albert Einstein.

Figure 1.8: Einstein and his second wife, Elsa.

References

Autschbach, J. (2014). "Relativistic calculations of magnetic resonance", *Phil. Trans. R. Soc. A* **372**, 20120, 489.

Bethe, H.A. (2000). "The German uranium project", *Phys. Today* **July**, 34.

Droin, C. *et al.* (2019). "Low-dimendional dynamics of two coupled biological oscillators", *Nature Physics* **15**, 1086.

Feynman, R.P. (1988). *QED-The Strange Theory of Light and Matter* (Princeton University Press).

Naumann, R. and Stroke, H. (1996). "Einstein and the atomic clock", *Phys. World* **April**, 76.

Parsons, P. (2011). *3-Minute Einstein* (Blume, Barcelona).

Taylor, S.R. *et al.* (2016). "Are we there yet? Time to detection of nanohertz gravitational waves based on pulsar-timing array limits", *Astrophys. J. Lett.* **819**, L6.

Wynands, R. and Weyers, S. (2005). "Atomic fountain clocks", *Metrologia* **42**(3).

Chapter 2

Invariant Light Velocity

2.1 Concepts of Light Propagation

What is light? And what is its velocity? These are questions that philosophers, theologians, and scientists have asked for centuries. Since the classical Greek epoch, many hypotheses had been circulating, and it was assumed that light was emanating from objects and human vision arose from the eyes to capture "replicates of the object".

Pitagoras (585–505 A.C.), e.g. thought that light was radiated by the eyes in straight lines. Our vision was then a result of the reflection by distant bodies, and eyes were at the same time a detector of light as a radar dish today.

In the 17th century, Newton elaborated his corpuscular theory of light, according to which light was a jet of particles originating from a lamp. By the early 18th century, it was widely believed that light was composed of small particles. Phenomena like reflection, refraction, and shadows of bodies could be expected from torrents of particles. Newton tried to explain refraction by postulating that the particles increased their velocity when the medium's density increased. The scientific community, aware of Newton's prestige, accepted his emanation theory.

However, Huygens, a contemporary of Newton, was inclined to a theory which describes light as waves propagating in space, and deduced the laws of reflection and refraction. He even explained the double refraction of calcite via his wave model.

Today, we know that light reveals wave–particle duality and that some phenomena, as the photoelectric effect of Hallwachs and Einstein, are easier to understand using massless photons and others, as interference, with the

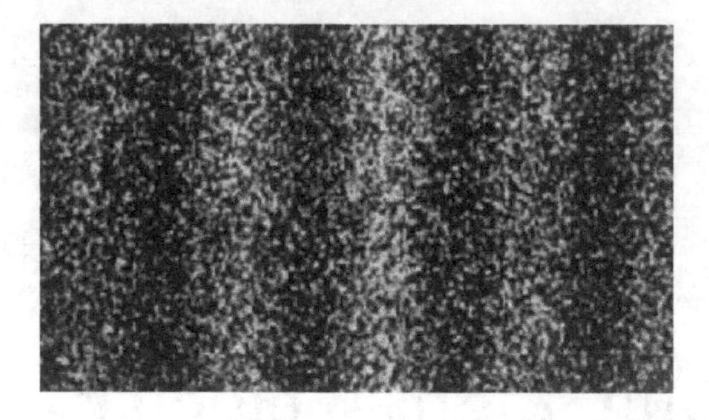

Figure 2.1: Interference created by a bunch of particles. In this case, 140,000 electrons emitted by an electron microscope (Tonomura, 2005) show both, the random impact and interference patterns of de Broglie type matter waves.

concept of wave. This duality is central to the Copenhagen interpretation of Quantum Mechanics.

At the time of Galileo (1564–1642), a popular belief about the propagation of light was that it instantly spread, and therefore, had an infinite velocity. Galileo doubted that the velocity of light would be infinite and proposed an experiment: Two people take lamps and place it on top of two mountains, one person turns his on and the other one must turn on the second as soon as he sees the light of the first person's lamp. This way one could calculate how much time had passed before the light on the other mountain was seen. But then, the velocity of light is so high that it impeded its detection via this kind of experimental set-up. Later on, measurements of the velocity of light were successfull as summarized in the following table:

Date	Researcher	Country	Procedure	Estimated velocity (km/s)
1676	Ole Rømer and J. Cassini	Denmark & Italy	In collaboration with Cassini, Rømer detected that the time between eclipses of Jupiter's satellite, Io, was less when the distance from Jupiter to the Earth decreased, and vice versa. He deduced that it was due to light having to propagate more when Earth is in opposition to Jupiter.	230 000

(*Continued*)

Date	Researcher	Country	Procedure	Estimated velocity (km/s)
1728	James Bradley	England	The velocity of light was estimated by observing stellar aberration, which is the apparent displacement of stars due to the movement of the Earth around the sun.	301 000
1849	Armand Fizeau	France	A light beam was reflected by a mirror 8 km away. The beam passed through a sprocket wheel whose angular velocity was increased until the returning beam completely passed the next gap.	315 000
1850	León Foucault	France	He improved Fizeau method by substituting sprockets by rotating mirrors.	298 000
1858	Bernhard Riemann	Germany	An invariant relativistic wave equation for the electric potential $\varphi = A^0$ was considered in an attempt to reconcile his scalar electrodynamics with the experiments of Kohlrausch and Weber (1855). The velocity of light is estimated based on the definition: $c \equiv 1/\sqrt{\varepsilon_0 \mu_0}$.	310 738
1891	René P. Blondlot	France	Stationary waves with nodes and antinodes spaced in regular distances are passing a pair of parallel wires. Knowing the eigenfrequencies and distances between the nodes, the velocity of radiation can be calculated.	297 600
1926	Albert A. Michelson	USA	With rotating mirrors, he measured the time that light took to make a round trip between the Wilson and San Antonio mountains in California.	299 520
1906	E. B. Rosa and N. E. Dorsey	USA	The capacitance C of a capacitor was measured in electromagnetic units and compared with C in electrostatic units.	299 781
1950	Louis Essen	England	He used radiation to produce stationary waves in a metal cylinder enclosing a small cavity.	299 792
1958	Keith D. Froome	England	Finally, use is made of a microwave interferometer and a Kerr cell.	299 792.5

Albert Abraham Michelson (1852–1931) and Edward W. Morley (1838–1923) conducted experiments with an interferometer that bears their names; they proved the invariance of the velocity of light, later on regarded as a Principle of Relativity of Einstein. In 1907, Michelson received the Nobel Prize in Physics for his work.

Currently, the velocity of light c is not measured anymore but has been established as a fixed constant in the International System of Units. Since 1983, the meter has been redefined as the length that light travels in a vacuum within the fraction $1/299\,792\,458$ of $1\,\mathrm{s}$, so that the velocity (*celerity*) of light is defined exactly as $c = 299\,792.458\,\mathrm{km/s}$.

2.2 Rømer and the Moons of Jupiter

The discovery of the finite velocity of light by the Danish astronomer Ole Rømer, in an article published in 1676, was based on the observation of eclipses of Jupiter's satellite Io. The orbits of Io around Jupiter are similar to those of the Moon around the Earth. The sun illuminates Jupiter, projecting its shadow in space. Since the satellite Io is almost in the plane of the orbit of Jupiter around the Sun, Io enters the shadow projected by Jupiter: Hidden during a time interval (being eclipsed), it continues its path around Jupiter.

Cassini was responsible for calculating the period of Io around Jupiter, using the time interval elapsed between two consecutive eclipses; he estimated an average duration of 42.5 h. However, after repeating the observation of the eclipses for several months, he realized that when Earth was further away from Jupiter, Io's period appeared larger than the average value. Rømer deduced that the reason for these differences is the finite velocity of light (which at that time was still considered infinite). Due to the variation of the distance between Jupiter and Earth, photons took longer to reach Earth when she was in opposition to Jupiter and vice versa (see Fig. 2.2).

Through the apparently varying periods of Io, Rømer could estimate the time light took to travel the diameter of Earth's almost circular orbit. The ratio between the diameter and the time light takes in traversing it is the velocity of light. Currently, knowing that the precise value of this time is about 1000 seconds and the diameter of Earth's orbit is 2 AU[1] $\simeq$ 300 million kms, we obtain $c \simeq 3 \times 10^8\,\mathrm{km}/1000\,\mathrm{s} = 3 \times 10^8\,\mathrm{m/s}$ (see Fig. 2.3).

[1]The astronomical unit has nowadays a fixed value of $1\,\mathrm{AU} = 149\,597\,870\,700\,\mathrm{m}$.

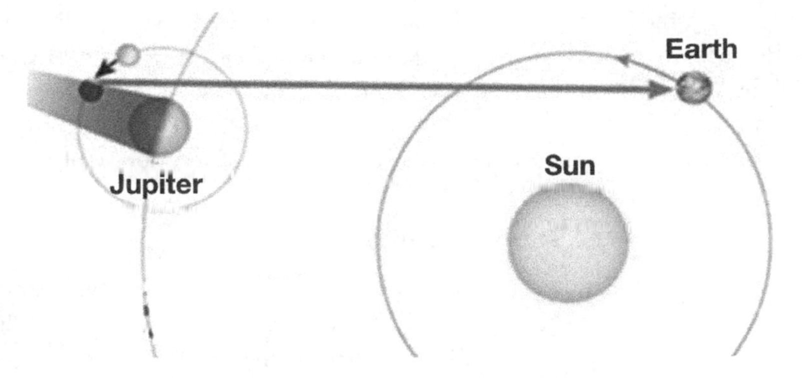

Figure 2.2: Eclipse of the satellite Io when the Earth is almost in opposition to Jupiter. An observer on Earth records the elapsed time between two consecutive eclipses.

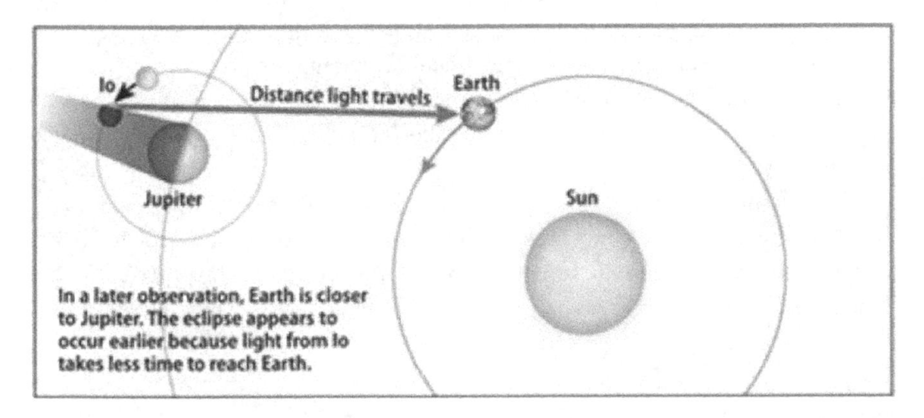

Figure 2.3: Eclipse of the satellite Io when the Earth is closer to Jupiter. An observer notes the time between two consecutive eclipses is less than that recorded when the Earth is in opposition to Jupiter. This variation is due to the time that photons take to cover the distance between the satellite and Earth.

Although the discovery of the finite velocity of light is attributed to Rømer, there is a text in the Paris Observatory which seems to confirm, the minority view, that Jean-Dominique Cassini observed the satellites of Jupiter and worked in the Paris Observatory with Rømer. First he proposed a "successive movement" (i.e. finite) of light and gave a rather good approximation of the additional time it would take to cross the diameter of the Earth's orbit arround the sun. However, doubts invaded Cassini, discarding the hypothesis of the finite velocity of light, while Rømer held it up and published it without taking into account the doubts of Cassini and other scientists. Some historians of science believe that it was a joint discovery, see Bobis & Lequeux (2008) for more details. When Christian Huygens read the article by Rømer, he accepted with enthusiasm the idea of the finite velocity of light and, using the data obtained by Cassini and Rømer, he deduced the numerical value of 230 000 km/s for the velocity of light.

2.3 Terrestrial Measurements of Fizeau

The first non-astronomical measurement of the velocity of light was accomplished in 1849 by the French physicist Hippolyte Fizeau. He placed a light source with a lens system on a hill and, approximately at 8.63 km, a second mirror on another hill (Fig. 2.4 shows the light source). Light emerged from the source, passing through a lens, whose purpose was to focus it on a half-plated mirror transmitting half of the light to the observer (using another lens) and the other half being reflected. Once the light passed through the hole in the sprocket of a rotating wheel, it went through a further lens forming parallel rays. Another lens focused the light on the mirror placed on the second hill. In that mirror, all the light was reflected and returned via the same optical path.

Fizeau noticed that, at lower angular velocities, no light was visible because the light that passed through one hole of the cogwheel was obstructed by the next tooth after being reflected by the mirror at the distant hill. Angular velocity is then increased such that light passed

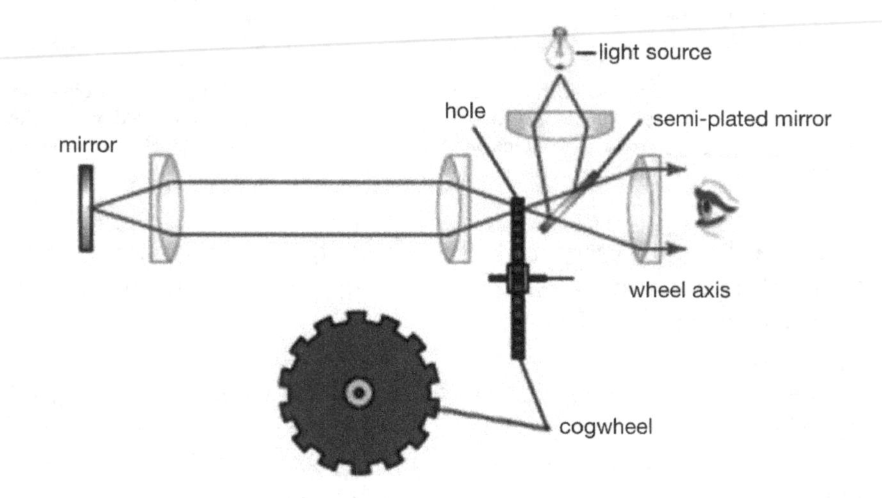

Figure 2.4: Experiment of Fizeau.

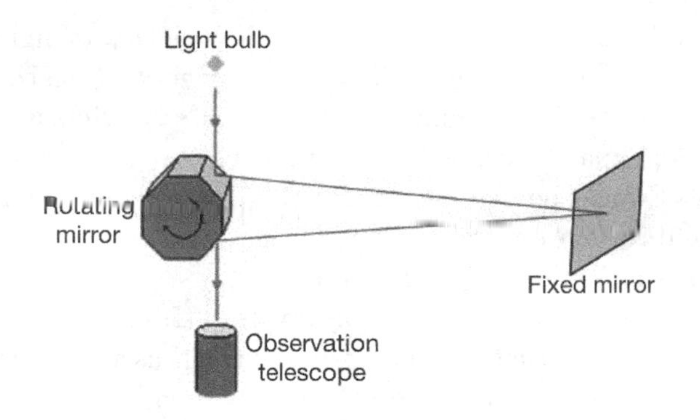

Figure 2.5: Foucault's experiment: The segmented mirror rotates one-eighth of its circumference during the time light takes to go to the fixed mirror and back. Then, the next face of the mirror is in the proper position to reflect light toward the observer.

through the next hole of the cogwheel. The time required, turning the angle between two successive holes, was equal to the time taken for light to travel the distance of the cogwheel to the mirror and its return. Thereby, a value of $c = 315\,000\,\mathrm{km/s}$ was found.

Based on Fizeau's ideas, Foucault developed later on a similar experiment replacing the cogwheel by a rotating mirror, obtaining more precise results (see Fig. 2.5).

2.4 The Michelson–Morley Interferometer

Maxwell's theory of electromagnetism implies that light is an ondulatory phenomenon and, like all waves, would need a medium to propagate. The hypothetical medium in which light propagates was named "Luminiferous aether".

The interferometer experiment, conducted in 1881 by Albert A. Michelson and 1886 together with Edward Morley, is considered as the first test against the aether hypothesis. The original purpose of Michelson and Morley, however, was to measure the velocity of the Earth relative to the center of the solar system.

Each year, the Earth travels a large distance on its orbit around the sun at a velocity of about $v_\oplus = 30\,\mathrm{km/s}$ (more than 100,000 km/h). It was believed that the direction of the "aether wind" relative to the position of the sun would vary when measured on Earth, and thus could be detected. For this reason, and to avoid the effects which the Sun may provoke while moving around the center of the Milky Way, the experiment should be carried out during various seasons of the year.

The experiment intended to measure the velocity of light on Earth, in different directions and different months of the year, since the relative movement of the "aether" cannot be equal at the same time in two different directions. An apparent change in the velocity of light was expected. The "aether wind" would be like the stream of a river, where a swimmer moves in the same direction or against the current. At times, the swimmer will be slowed down and otherwise driven forward.

In the basement of the Potsdam observatory (Bleyer *et al.*, 1979) at sea level, Michelson constructed what is now known as the Michelson interferometer. It consists of a lens and a semi-plated mirror, which divides (monochromatic) light into two beams of coherent light (now more easily realizable by a laser). This was achieved by simultaneously sending two beams of light (from the same source) in perpendicular directions, let them travel equal distances (or equal optical paths) reaching a point in common. There an interference pattern is created that depends on the velocity of light in the two arms of the interferometer. Any difference in this velocity caused by the different direction of propagation of light (relative to the "aether") would be detected. The journey time interval for a round trip along arm two is

$$\Delta t_{\mathrm{arm2}} = \frac{L}{c+v} + \frac{L}{c-v} = \frac{2Lc}{c^2 - v^2} = \frac{2L}{c}\left(1 - \frac{v^2}{c^2}\right)^{-1}.$$

For arm one, due to the movement of the semi-plated mirror (see Fig. 2.6), the Pythagorean theorem implies:

$$\Delta t_{\mathrm{arm1}} = \frac{2L}{(c^2 - v^2)^{1/2}} = \frac{2L}{c}\left(1 - \frac{v^2}{c^2}\right)^{-1/2}.$$

Consequently, the difference in time travel between arm one and two is

$$\Delta t = \Delta t_{\mathrm{arm2}} - \Delta t_{\mathrm{arm1}} = \frac{2L}{c}\left[\left(1 - \frac{v^2}{c^2}\right)^{-1} - \left(1 - \frac{v^2}{c^2}\right)^{-1/2}\right].$$

Because $v^2/c^2 \ll 1$, we can simplify this expression using the binomial formula $(1-x)^n \approx 1-nx$ for $x = v^2/c^2$. In our case, we find that $\Delta t \approx Lv^2/c^3$. When the interferometer is rotated 90° in a horizontal plane, there results a doubling of the time difference:

$$\Delta d = c(2\Delta t) = \frac{2Lv^2}{c^2}.$$

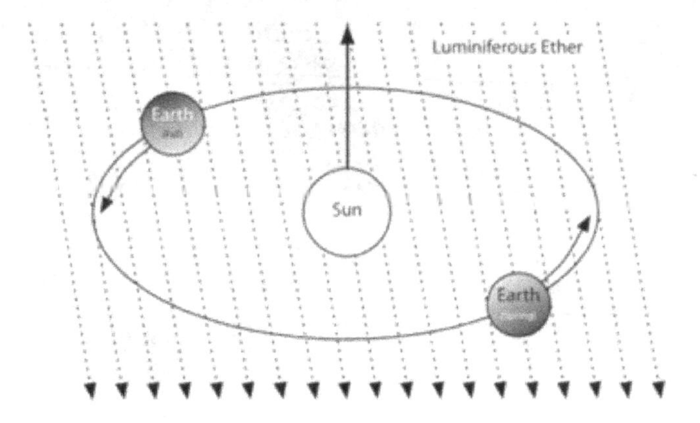

Figure 2.6: Movement of the Earth with respect to a hypothetical aether "wind".

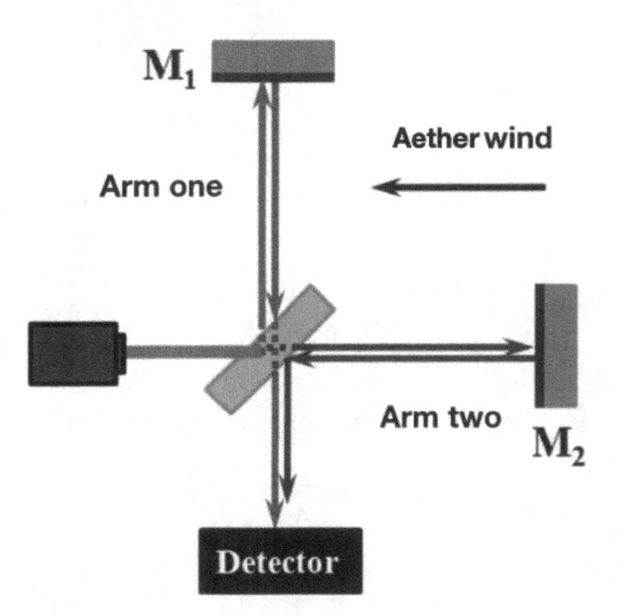

Figure 2.7: The light is split into two perpendicular beams (red and blue). After being reflected by the mirrors M_1 and M_2, they interfere in the detector. Since the velocity of the "aether" would change the trajectory of one of the beams, one would expect to find a difference in the optical path length.

Later on, in experiments conducted by Michelson and Morley, the arms had length L, approximately 11 m due to "double" or multiple light paths (see Fig. 2.7). Using the tangential velocity $v_\oplus = 3.0 \times 10^4$m/s of the Earth around the Sun, there would result a time difference spent in the arms.

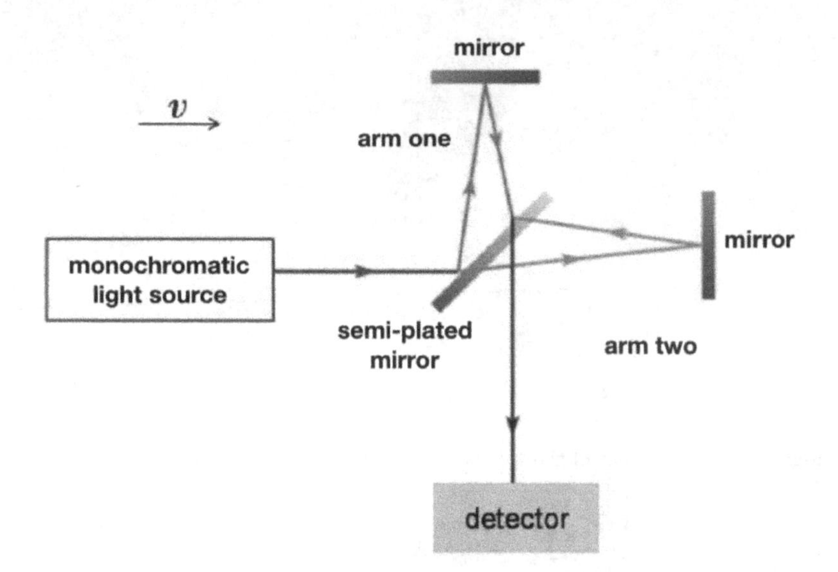

Figure 2.8: Optical path (more realistic).

This extra travel distance should produce a noticeable change in the fringe pattern:

$$\Delta d = \frac{2(11\,\text{m})(3.0 \times 10^4\,\text{m/s})^2}{(3.0 \times 10^8\,\text{m/s})^2} = 2.2 \times 10^{-7}\,\text{m}.$$

A change in the path length of a wave would correspond to a change of fringes. It is equal to the difference of the path divided by the wavelength:

$$\text{change} = \frac{\Delta d}{\lambda} = \frac{2Lv^2}{\lambda c^2}.$$

If we use green (500 nm) light, waiting for a rotation of 90° of the interferometer due to the trajectory of the Earth, the relative change of the fringes is

$$\frac{\Delta d}{\lambda} = \frac{2.2 \times 10^{-7}\,\text{m}}{5.0 \times 10^{-7}\,\text{m}} \approx 0.44.$$

The instrument used by Michelson and Morley could detect changes in the fringes of only 0.01 and had an observational error of 25%. This was reduced to 1% in the repetition of Morley and Miller of their experiment in 1904, corresponding to a predicted error $\Delta c/c \simeq 10^{-6}$ for the propagation

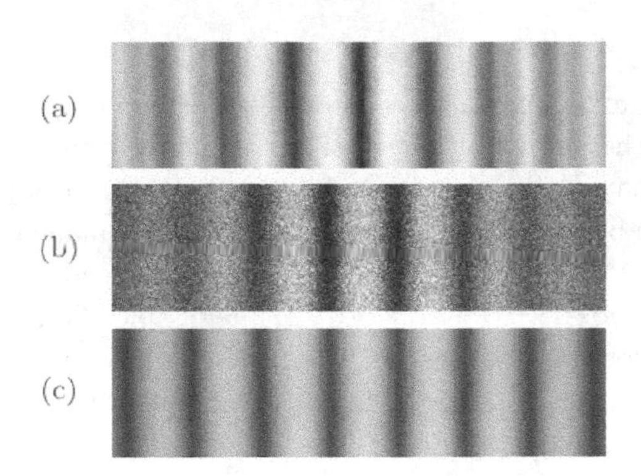

Figure 2.9: Shift of interference patterns.

of light. However, they did not detect any change in the fringe pattern. Therefore, it was concluded that there is no "aether" or absolute frame.

2.5 Decisive Lorentz Contraction

After the ad hoc proposals of FitzGerald, Einstein solved the problem in 1905 within his Special Relativity, which predicts a contraction factor $\gamma = 1/\sqrt{1 - v^2/c^2}$ introduced already by Lorentz:

$$\text{Photons} \perp \vec{v}_\oplus \qquad \Delta t_{\text{arm1}} = \frac{2L}{c}\left(1 - v^2/c^2\right)^{-1/2},$$

$$\text{Photons} \parallel \vec{v}_\oplus \qquad \Delta t_{\text{arm2}} = \frac{2L}{c}\left(1 - v^2/c^2\right)^{-1}.$$

The Michelson interferometer tells us experimentally that

$$\Delta t = \Delta t_{\text{arm2}} - \Delta t_{\text{arm1}} \simeq 0.$$

If arm two undergoes a Lorentz contraction $L \to L\sqrt{1 - v^2/c^2}$, then

$$\begin{aligned}\Delta t_{\text{arm2}} &= \frac{2L}{c}\left(1 - \frac{v^2}{c^2}\right)^{-1} \to \frac{2L\sqrt{1 - v^2/c^2}}{c}\left(1 - \frac{v^2}{c^2}\right)^{-1} \\ &= \frac{2L}{c}\left(1 - \frac{v^2}{c^2}\right)^{-1/2} = \Delta t_{\text{arm1}}.\end{aligned}$$

Accordingly, in Special Relativity the time difference is exactly $\Delta t_{\text{SR}} = 0$, and hence there is no change in the pattern of fringes.

2.6 Isotropy of Light

More recently, experiments have been conducted to test the isotropy of light (i.e. whether or not the velocity of light is the same in all directions) by comparing the resonance frequencies of two optical orthogonal resonators of Fabry–Perot type within a single base made of molten silicon oxide (quartz), see Fig. 2.10, rotated an on air-suspended plate, where the relativistic Sagnac effect (Pascoli, 2017) is negligible. Gyrolasers that monitor angular velocities with respect to an inertial frame, exploit this effect. An analysis of the data recorded during a complete year sets a limit for a possible light anisotropy:

$$\Delta c/c \simeq 1 \times 10^{-17}.$$

This is the most accurate laboratory test of the isotropy of c today, which allows to put limits on other parameters of the Standard Model of particle physics again on a level of 10^{-17}. See Herrmann *et al.* (2009).

2.6.1 *Comparison with the velocity of neutrinos and gravitational waves*

In 1987, the blue supergiant Sanduleak exploded as the supernova SN 1987 A. In the south, it was visible to the naked eye 383 years after the SN of Kepler. Despite an observed prompt (within 3 h) neutrino emission, the formation of a neutron star or another compact object is still difficult to confirm due to dust (Cigan *et al.*, 2019).

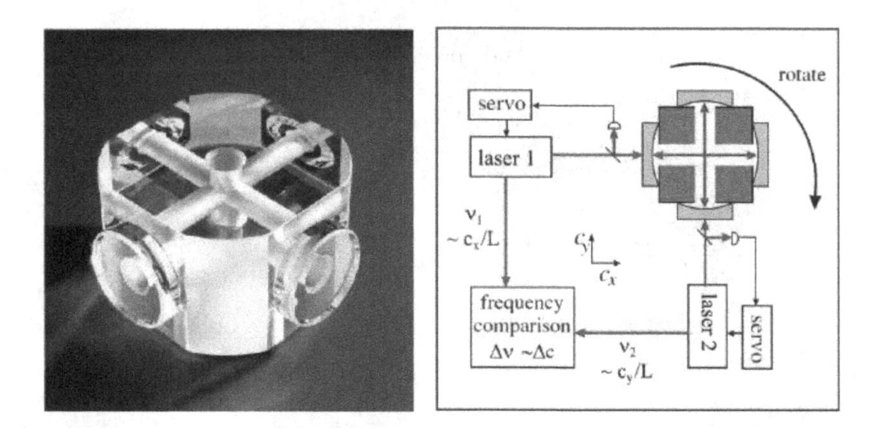

Figure 2.10: Interferometer in a single molten quartz block.

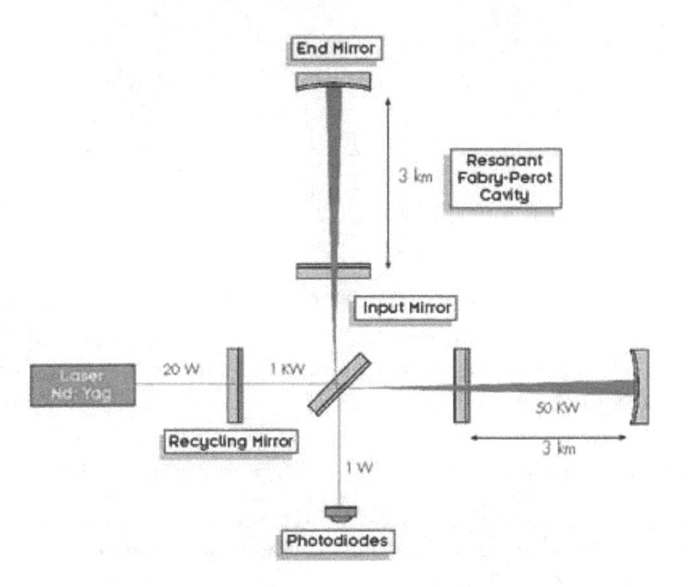

Figure 2.11: Virgo interferometer of the Fabry–Perot type with 3 km-long arms. Detectors, such as LIGO and Virgo, consist of a Michelson interferometer with a Fabry–Pérot cavity with several kilometers in both arms. This cavity stores photons for almost a millisecond while bouncing up and down between the mirrors. This increases the time a GW can produce geodesic deviations of the mirrors as test masses, resulting in a better sensitivity at low frequencies.

Nevertheless, the event led to the first estimate of the velocity v_ν of (electron) neutrinos on a cosmic scale. The comparison with the velocity of light c provided the estimate

$$(c - v_\nu)/v_\nu < 3hc/D \approx 2 \times 10^{-9}.$$

Although, neutrinos are experimentally shown to have a tiny mass, they travel with almost the same velocity as photons on a galactic distance of $D = 51\,\mathrm{kp}$ from Earth.

According to Einstein's General Relativity,[2] linearized gravitational waves (GWs) propagate with the same velocity as light in vacuum, i.e. $c_g = c$, if there is no dispersion: At LIGO (Laser Interferometer Gravitational Wave Observatory), a similar but more sophisticated Michelson

[2]GR relates mass (energy–momentum) and spacetime curvature in a similar way that Hook's law relates forces and spring deformation. Spacetime acts as a kind of an elastic medium and, if a mass distribution (billard balls) moves in an asymmetric way, then (solitonic) ripples in spacetime travel outwards as transversal gravitational waves.

interferometer with 4 km arm length was constructed to detect GWs generated by inspiraling neutron stars. These GWs have reached LIGO with the same velocity c of photons (gamma rays) except for a tiny time difference of less than 3 s (Boran *et al.*, 2018) on a cosmic scale of 130 million light-years.

2.7 Superluminal Velocities?

Suppose there would exist a signal with $v = \infty$ (or $v > c$). An experiment uses this signal or superluminal wave at point A. A mirror in movement with $v < c$ reflects this signal to point C, which could be constructed by the axis parallel to x'. Since $t_C < t_A$, there would be a problem with the causality for superluminal signals or tachyons. According to Saint Agustin, only your "Guardian Angel" can be in different places at the same time:

> "Et sic angelus in uno instanti potest esse in uno loco, et in alio instanti in alio loco, nullo tempore intermedio existente"

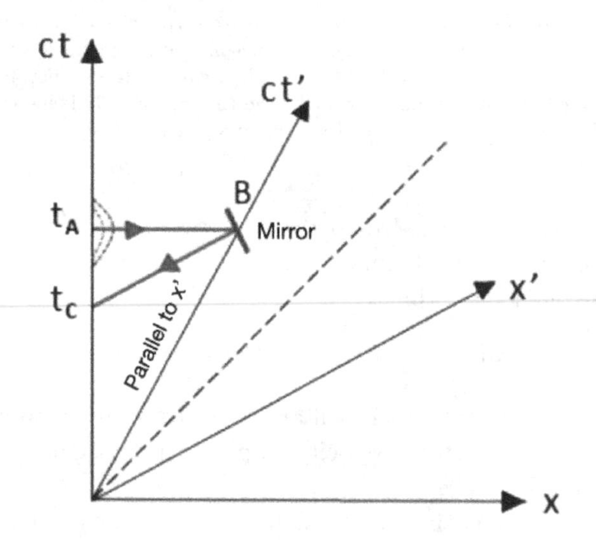

In experiments called "lunar ranging" (Williams *et al.*, 2012), which consist of sending a beam of laser light to the moon by a telescope of large aperture, time is measured for reaching the moon. Geometrically, this beam seems to be "rotating" at an angular velocity apparently "superluminal", see Fig. 2.12. But only without considering the time of about one second that the photons take to reach the retro-reflector placed on the Moon, and the time back.

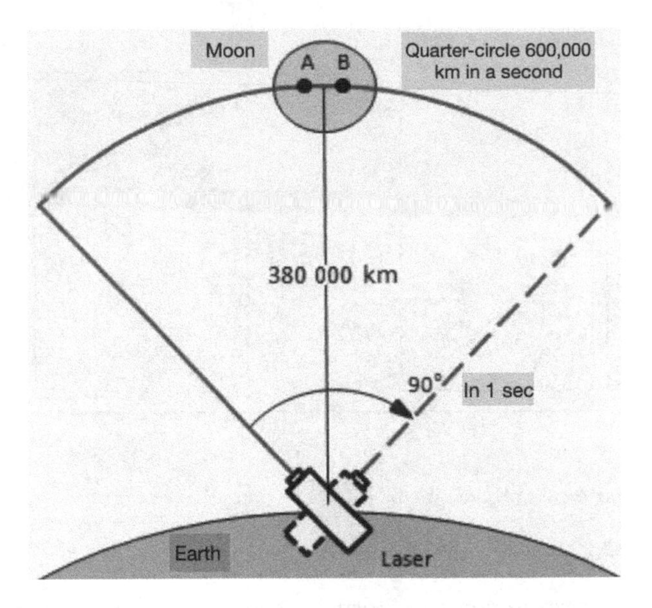

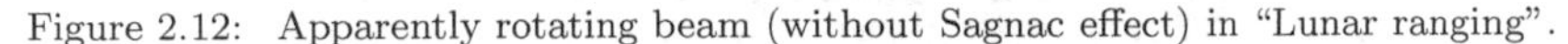

Figure 2.12: Apparently rotating beam (without Sagnac effect) in "Lunar ranging".

2.7.1 *Tunnel effect*

Recently, signals carried by evanescent modes have been investigated whether or not they are traveling faster than light. To compare, microwave pulses are passing through the vacuum and the evanescent medium realized by a constrained wave guide: The result is that the signal in the tunnel has run its course faster than the vacuum's signal. Even more surprising is that the travel time of the signal in the tunnel does not seem to depend on the length L of this barrier.

This effect is known as "tunnel effect" in quantum mechanics. In microwave guides,[3] this has caused a debate (Nimtz, 2011) about "superluminal" signals. However, the tunnel, in the frame of photons, almost disappears due to the Lorentz contraction, since:

$$L \to L\sqrt{1 - v^2/c^2} \simeq 0,$$

For more details, see Mielke & Marquina (2013).

---

[3]A waveguide is any physical structure that guides electromagnetic waves. The first waveguide was proposed by Joseph John Thomson in 1893 and experimentally realized by O.J. Lodge in 1894.

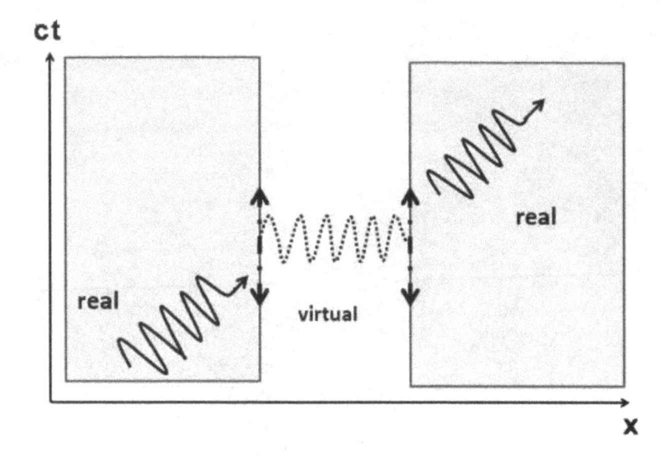

Figure 2.13: Minkowski diagram for photons in a microwave guide.

2.8 Homework

1. Use a microwave oven to determine the velocity of light: Remove the turntable and replace it with a cardboard of soft candies (for example, little marshmallows). Bake for a few minutes: Then you can see slight depressions in the layer of candies, which allow you to measure the wave length λ (or $\lambda/2$ of stationary waves). Since the frequency f is fixed in the GHz range, the velocity of the microwaves in the air (which is almost equal to $c = \lambda f$ in vacuum) is obtained just by multiplication (Vollmer, 2004).
2. Realize Thomas Young's double-slit experiment of 1801 via a CD. Use a laser pointer to reflect its (coherent) light and photograph the resulting interference pattern on a wall.

References

Boran, S. *et al.* (2018). "GW170817 falsifies dark matter emulators", *Phys. Rev. D* **97**, 041501.

Cigan, P. *et al.* (2019). "High angular resolution ALMA images of dust and molecules in the SN 1987 A ejecta", *Astrophys. J.* **886**, 51.

Pascoli, G. (2017). "The Sagnac effect and its interpretation by Paul Langevin", *Comptes Rendus Physique* **18**, 563.

Van den Brand, J. (2019). "A primer on LIGO and Virgo gravitational wave detection", European School of High-Energy Physics, St. Petersburg, September 16, 2019.

Chapter 3

Time Dilation

3.1 Idealized Einstein Clock

As an example of Einstein's famous "thought experiments" that explain essentials of Special Relativity is the "light clock", mentioned for the first time by Lewis & Tolman (1909). Nowadays, this clock can be made with a laser, a mirror, and a photon detector. Its "proper time" is $\Delta\tau = 2L/c$, where c is the invariant velocity of light (see Fig. 3.1).

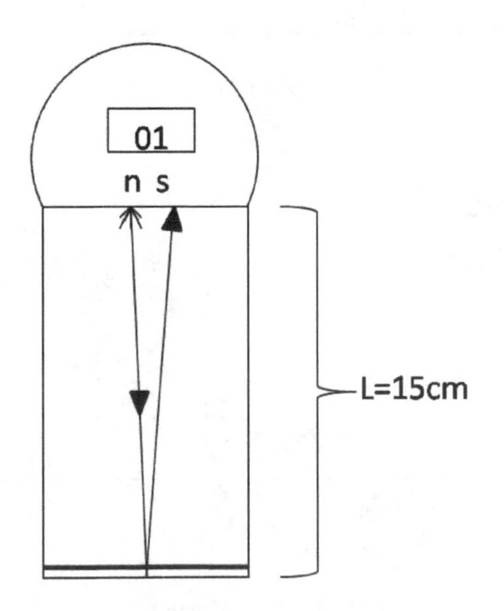

Figure 3.1: "Einstein clock" made with a resonating laser. Light reflection provides nanosecond "tics".

For example, for a "table clock" with length $L = 15$ cm, we have a proper time of $\Delta\tau = 2L/c \simeq 10^{-9}$s $= 1$ ns for one tic. As mentioned earlier, the accuracy of a Caesium atomic clock is around $\Delta\tau/\tau = 4\times 10^{-16}$.

3.1.1 *Comparison of light clocks in different inertial frames*

I: Initially, two clocks at rest, A and B, are synchronized (both at 10 ns initially, see Fig. 3.2).

I′: The clock C has the velocity v, relative to the frame of reference I (see Fig. 3.3).

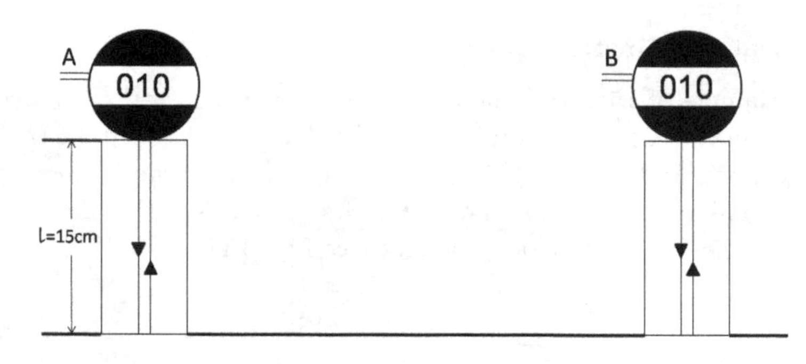

Figure 3.2: Frame I.

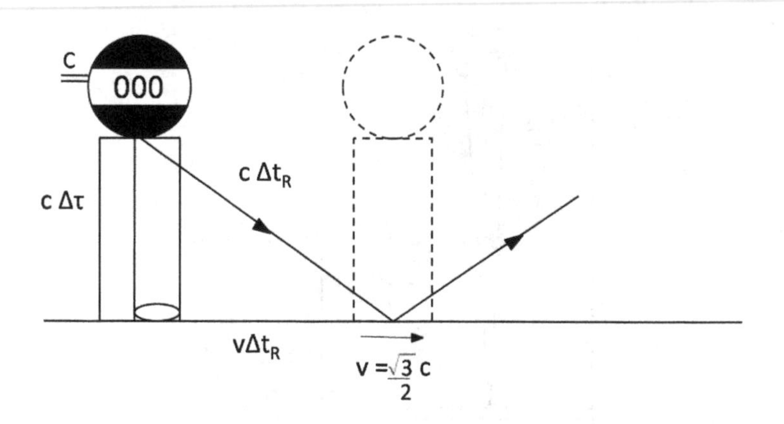

Figure 3.3: Frame I′.

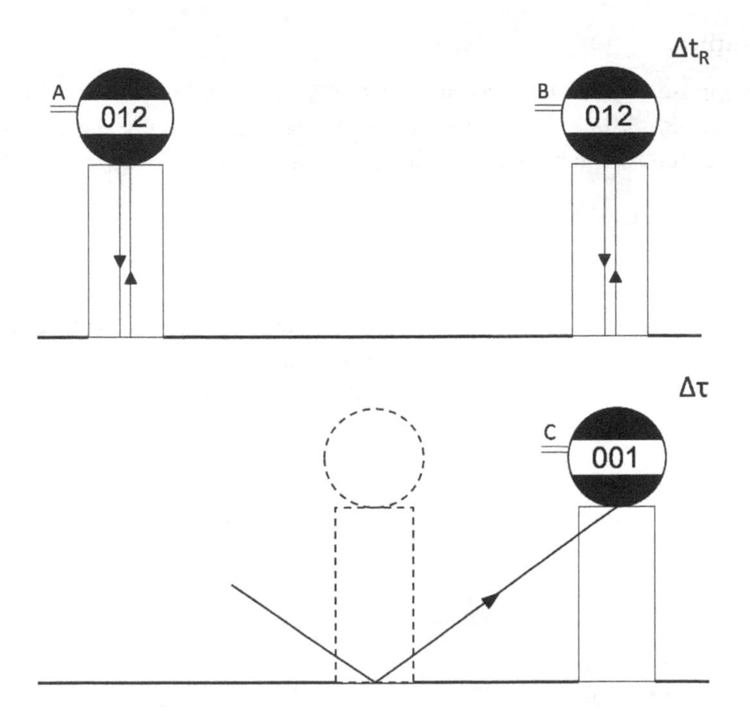

The Pythagorean theorem[1] for distances in the rectangular triangle traveled by light (in vacuum) tells us that

$$(c\Delta\tau)^2 + (v\Delta t_{\mathrm{R}})^2 = (c\Delta t_{\mathrm{R}})^2.$$

Clearing $\Delta\tau$, the **time dilation** is

$$\Delta\tau = \Delta t_{\mathrm{R}}\sqrt{1-(v/c)^2} \Rightarrow \Delta t_{\mathrm{R}} = \frac{\Delta\tau}{\sqrt{1-(v/c)^2}} \geq \Delta\tau.$$

Historically, this formula already appears in a work by Joseph Larmor in 1897, see Kittel (1974), together with the following Lorentz factor:

$$\gamma = 1/\sqrt{1-(v/c)^2} \simeq 1 + \frac{1}{2}(v/c)^2 \quad \text{for } v \ll c.$$

Time dilation is a real effect of physics, in as much as many clocks are built at the microscopic level with electromagnetic interactions, i.e. with "virtual" photons.

[1]This theorem had previously been applied by Babylonian "farmers" to regain their ground after a flooding caused by a river. On the other hand, Pythagoras from Samos founded a school to spread this ancient geometric knowledge. Integers such as $m^2 + n^2 = k^2$ were found by Fibonacci.

Examples:

(a) Suppose clock C has a relative velocity of $v = \sqrt{3}c/2$. Considering, for example, $\Delta t_{\mathrm{R}} = 1\,\mathrm{ns}$ in one of the triangles and taking the doubling of time due to the reflection of light into account, its proper time is

$$\Delta\tau = 2\left(1\,\mathrm{ns}\sqrt{1-\frac{3}{4}}\right) = 1\,\mathrm{ns}$$

as expected.

(b) A pion disintegrates in a time interval of $\Delta\tau = 26.0$ ns; corresponding to its proper time. Let the pion be in motion with the velocity $v = 0.913c$ relative to the Earth's surface. Calculate the time interval of decay in the (Earth's) rest frame.

Solution:

$$\Delta t_{\mathrm{R}} = \frac{\Delta\tau}{\sqrt{1-(v/c)^2}} = \frac{26.0\,\mathrm{ns}}{\sqrt{1-(0.913)^2}} = 63.7\,\mathrm{ns}.$$

3.2 Time Dilation of Muons

The muon μ is an elementary particle like the electron, with no internal structure and belongs to the lepton family. Its mass is more than 200 times the mass of the electron, more precisely being $m_\mu = 206m_e$, and it is unstable, so it decays rapidly. Its decay usually generates an electron, a neutrino, and an antineutrino:

$$\mu \longrightarrow e^- + \nu_\mu + \bar{\nu}_e.$$

By the law of radioactivity, their initial number N decays exponentially as

$$N = N_0 e^{-\lambda t} = N_0\left(\frac{1}{2}\right)^{t/\tau}.$$

We can use its decay constant $\tau = 1.52\,\mu\mathrm{s}$ as the proper time on the scale of microseconds (see Fig. 3.4).

In an experiment conducted at Conseil Européen pour la Recherche Nucléaire (CERN), Geneva, the muons had a circular velocity of $v = 0.99942c$, that is, almost the velocity of light. If we compare the laws of radioactivity in the circular movement with that at rest, it turns out that

$$t_{\mathrm{R}} = \frac{\tau}{\sqrt{1-(v/c)^2}} = 29.4 \times \tau = 44.6\,\mu\mathrm{s},$$

where t_{R} is the time in the rest frame.

Experiments confirm with 0.1% Special Relativity, the diminutive effect of centripetal acceleration can be ignored (see Fig. 3.5).

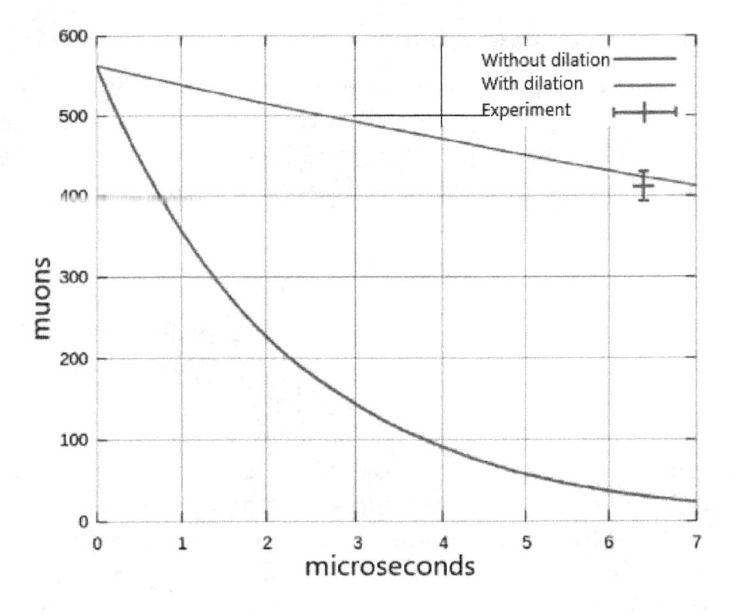

Figure 3.4: Number of muons with and without time dilation in μs.

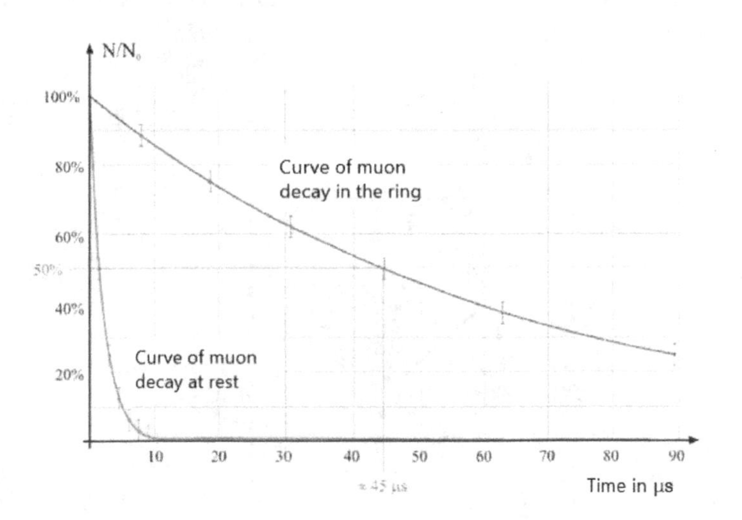

Figure 3.5: Comparison of the decay time of muons.

3.3 Gravitational Time Dilation

When a clock moves quickly, relative to an observer, the interval between each tic is different (longer) compared to the one measured when the clock was at rest in an inertial frame. In Special Relativity, this is the cinematic time dilation

$$dt = \frac{d\tau}{\sqrt{1 - v^2/c^2}} = \gamma d\tau \geq d\tau.$$

Doing a binomial expansion and considering only small velocities, we have

$$dt \simeq \left(1 + \frac{v^2}{2c^2}\right) d\tau.$$

Something similar is expected to occur when the clock moves at variable velocity. On the other hand, according to the principle of equivalence, the effects produced by gravitation are locally the same as those produced by acceleration. Therefore, the mere presence of gravitational masses in the proximity of a clock will also cause the clock to march slowly, even if the observer does not move toward it. This is the gravitational time dilation. Einstein suggested an imaginary experiment in which the value of this dilation can be measured, for a weak gravitational field such as that of the Earth (see Fig. 3.6).

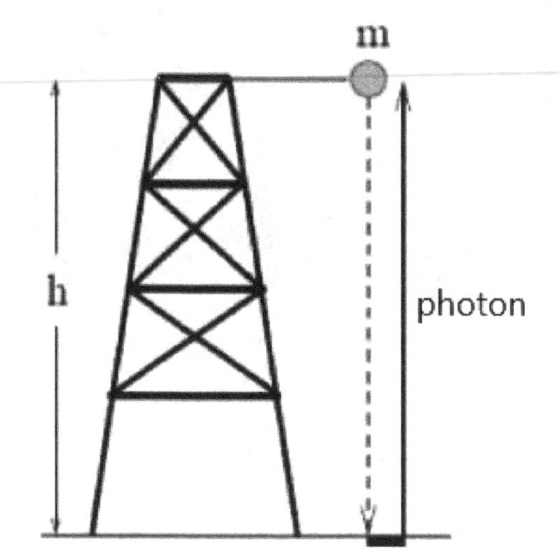

Figure 3.6: Gravitational time dilation experiment.

A mass m is dropped from the top of a tower of height h onto the surface of the Earth. When it reaches the ground, its velocity will be, according to Newton, $v = \sqrt{2gh}$ and therefore its rest energy, $E_0 = mc^2$, will have increased by the corresponding amount due to the kinetic energy acquired. Thus, on the ground, a rough, non-relativistic expression is $E \simeq E_0 + \frac{1}{2}mv^2 = mc^2 + mgh$. Suppose that all this energy converts to a photon with $E = hf$ that is emitted upwards. By energy conservation, the photon's energy when it reaches the top of the tower must be $E_0 = hf_0$, i.e. its frequency f will have decreased: A photon escaping a gravitational potential suffers from red-shifts. So,

$$\frac{E_0}{E} = \frac{hf_0}{hf} = \frac{mc^2}{mc^2 + mgh} = \frac{1}{1 + gh/c^2} \approx 1 - \frac{gh}{c^2}$$

or, after canceling Planck's constant,

$$\frac{\Delta f}{f_0} = \frac{f - f_0}{f_0} = \frac{gh}{c^2}.$$

This experiment is idealized, but a sophisticated version of it was accomplished by Pound & Rebka (1960). The emission of an atomic transition must be red-shifted by a fraction of 2.46×10^{-15} after ascending the 22.6 m of the stairway inside the Jefferson Physical Laboratory of Harvard University. This tiny difference could be measured thanks to the Mössbauer effect for gamma rays and the prediction was verified with an accuracy of 1%. This change in frequencies must be the same as that experienced by the tics of a clock. Remember that the time between two tics (period) is the inverse of the frequency. Therefore, we infer that time passes more slowly the more intense the gravitational potential is. So, if dt is the time interval between events measured at a height h above the surface of the Earth and $d\tau$ is the one measured at the ground level, we have

$$dt \cong \left(1 + \frac{gh}{c^2}\right) d\tau.$$

In General Relativity, the same approximation can be inferred from Schwarzschild's exact solution of Einstein's equations for the gravitational field of the planet Earth.

3.3.1 *Hafele and Keating experiment*

In the experiment of Hafele & Keating (1972), the behavior of atomic clocks aboard commercial aircraft were measured. Cinematic and gravitational

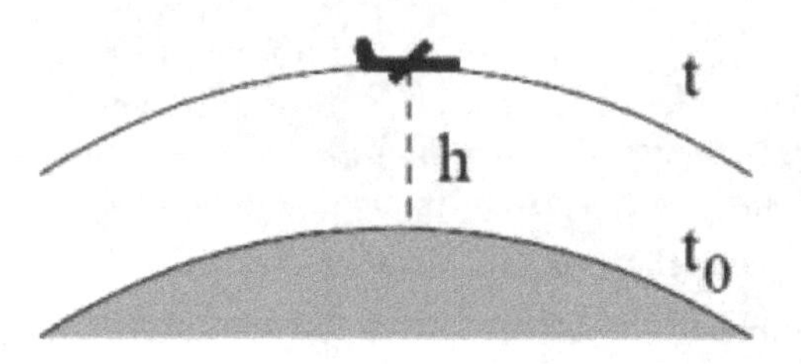

Figure 3.7: Time dilation according to the Hafele and Keating experiment.

time dilation were combined: The first is due to the relative movement between the in-flight clocks and the reference clock in Washington, which moves relative to the center of the Earth (locally being an inertial system freely fallen toward the sun). There is a delay of the clocks flying eastwards and the advancement of those flying westward. A second effect arises because the gravitational potential for on-board clocks is less than for the one left on the ground. This translates into an additional dilation, which we are already in a position to explicitly deduce. The result confirms Special Relativity predictions with 10% accuracy (see Fig. 3.7).

In 1976, the Smithsonian Astrophysical Observatory launched a Scout rocket to a height of 10 000 km. At that height, a clock should advance 4.25 parts of 1 000 faster than at ground level. During two hours of free-fall, the rocket was transmitting pulses from a maser oscillator acting like a clock, which was compared to the pulses of another similar ground clock. The result confirmed the gravitational time dilation at 0.02% (Vessot *et al.*, 1980), its best experimental determination to date.

3.4 The Apparent Twin Paradox

This fictional story begins on a New Year's Day when an astronaut (A) travels from Earth to the nearest space station, about 3 light-years away. Astronaut A travels at a constant velocity of $0.6c$ and once he has reached the spaceship, he stops and returns at the same velocity. When he returns, it is again new year and exactly 10 years have passed on Earth. This astronaut has a twin brother (B) who stays on Earth all that time (see Fig. 3.8).

Astronaut A returns 8 years older, while the twin, who stayed, is 10 years older, due to the dilation of time:

$$10\sqrt{1 - v^2/c^2} = 10\sqrt{0.64} = 8.$$

The paradox would be this: "From the point of view of astronaut A, Earth's clocks are slower, so A should be older when he comes back to Earth

Figure 3.8: Twins before and after traveling.

than his brother, not younger. This is a logical contradiction, therefore relativity is inconsistent."

Observers in the ship's reference system during departure agree that Earth's clocks are marching slower compared to them, 3.2 years compared to 4. In this moving frame, according to the *relativity of simultaneity*, the event that is simultaneous with event P (the "turn" of A) is event X.

When he arrives at the space station, he "mounts" to another frame of reference, now back, which also travels with $0.6c$ and again they agree that Earth's clocks go slower, 3.2 years compared to 4. In this return frame, the event that is simultaneous with P, according to the relativity of simultaneity, is the Y event, 3.2 years before returning, cf. Fig. 3.9.

But the analysis done with only these two frames of reference fails to count the 3.6 years between event X and Y, which is why A should be older than its twin according to the paradox. However, this is not a paradox, it is just an evaluation error. With the knowledge of *relativity of simultaneity*, astronaut A could easily realize that the gap between the two lines corresponding to his change (the turn) is 3.6 years. Alternatively, A could consult different observers in an (infinite) sequence of reference frames corresponding to the different velocities that the ship goes through during the turn and add up the different time increments on the Earth's clock: thus counting the missing 3.6 years.

At times, it is claimed that the resolution of this paradox involves General Relativity, as the traveling twin slows down at P and (de-)acceleration is equivalent to gravitation. However, acceleration plays no other role than providing asymmetry (as Von Laue explained already in 1913), whereby the

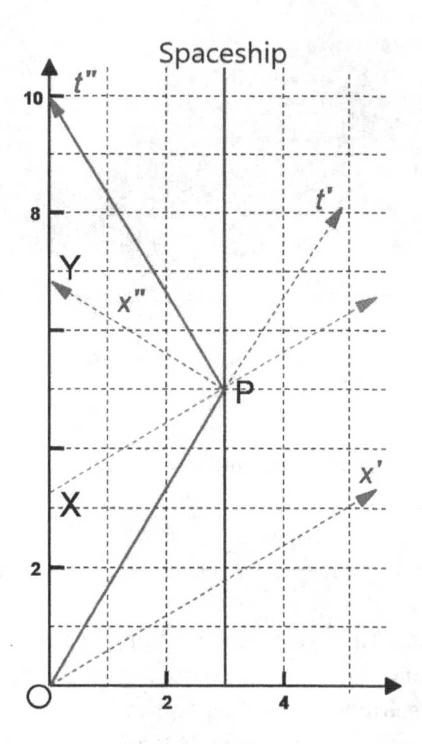

Figure 3.9: "Twin Paradox" from the point of view of the Earth bound twin. The Minkowski diagram indicates the axes (t', $\vec{x}'$) and (t'', $\vec{x}''$) of the astronaut's inertial frames with the velocity $|\vec{v}| = 0.6c$ relative to the Earth (Will, 2006).

traveller occupies more than one inertial frame, while the twin on Earth occupies only one. Relativity of simultaneity is the key, not gravity.

3.5 Time Dilation Measured by Optical Clocks

Moving observers or in varying gravitational potentials measure different "tics" on their clocks. These predictions of relativity had previously been observed with atomic clocks moving at rather high relative velocities, separately, with clocks at large differences of height. Afterwards, small changes $\delta f/f < 10^{-6}$ in the frequencies of the clocks could be observed by Mössbauer spectroscopy of γ-rays, using atomic interferometry.

However, it is now possible to observe the time dilation from relative velocities of more than 10 m/s (approximately the world record for men running 100 m like Usain Bolt) and from differences in height of 0.33 m. Recently, the frequency of two equal optical clocks in an "electromagnetic

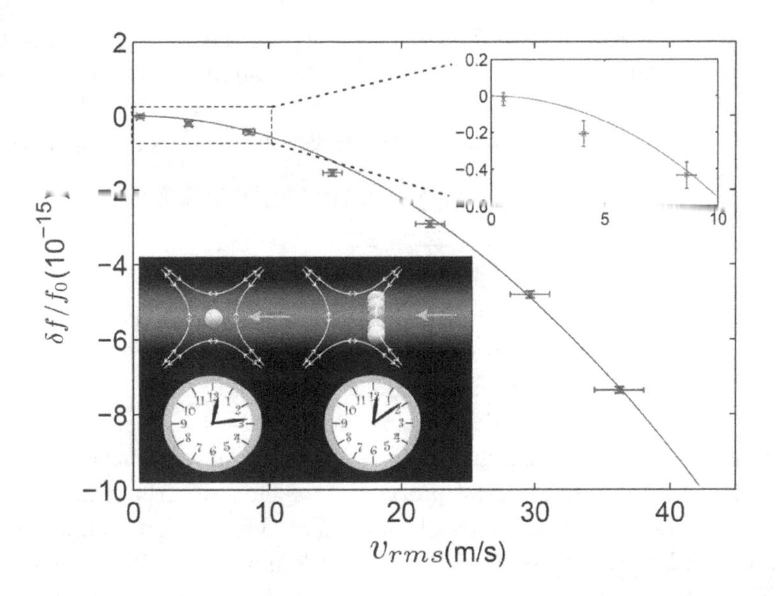

Figure 3.10: Time dilation at familiar car velocities (10 m/s = 36 km/h). The Al^+ ion, as one of the twin clocks, is displaced from the center of the confinement and subjected to an electric field (white field lines); causing a harmonic motion and the dilation of time. The fractional difference in frequency between the moving clock and the stationary clock is plotted against the mean velocity $v_{\rm rms} = \sqrt{\langle v^2 \rangle}$ of the moving clock. The continuous curve represents the theoretical prediction (Chou *et al.*, 2010).

trap" for Al^+ ions was compared with systematic uncertainties for the frequency of $\delta f/f \approx 10^{-17}$. When the comparison was made by 75 m of optical fiber, one of the atoms underwent a harmonic motion at a velocity of approximately 10 m/s, while other ionized atoms remained at rest. The accuracy and sensitivity of these Al^+ clocks, in view of the variations in frequency below 10^{-16} due to the dilation of time, were observed (see Fig. 3.10).

As an analogy to the "twin paradox", the Al^+'s ion is the traveling twin and its harmonic motion, in the trap, reaches several round trips. Time dilation causes a fractional change

$$\frac{\delta f}{f_0} = \frac{1}{\langle \gamma(1 - v_{\|}/c) \rangle} - 1,$$

in the frequency of the clock, where f is the ion's own resonance frequency, $v_{\|}$ is the parallel component of the velocity of the ion, and γ is the Lorentz factor.

Another consequence of Einstein's theory is that clocks advance more slowly near massive objects. Differences in the Earth's gravitational potential can be detected by comparing the instantaneous velocity of both clocks. For small changes of distance above the Earth's surface, a clock that is higher by Δh moves faster:

$$\frac{\delta f}{f_0} = \frac{g\Delta h}{c^2}.$$

Here, $g \approx 9.80\,\mathrm{m/s^2}$ is the local acceleration due to the gravitational potential of the Earth. Gravitational change provides a tiny time dilation of 1.1×10^{-16}s for each meter in height.

The comparison of atomic clocks is fundamental for international timekeeping, global positioning, and tests of fundamental physics. Optical-fiber links allow to compare the most precise optical clocks, without degradation, over distances up to thousands of kilometers. Very long baseline interferometry (VLBI) connected to H-masers, although originally developed for radio astronomy and geodesy, can overcome this limit and compare remote clocks through the observation of extragalactic radio sources, like quasars (quasi-stellar radio sources). By comparing clocks on different continents, in future, gravity potentials at these locations can be measured, cf. Haas (2011).

In Fig. 3.11(a) indicates the gravitational time dilation depending on the gradient $\vec{g}$, whereas Fig. 3.11(b) represents the fractional difference in

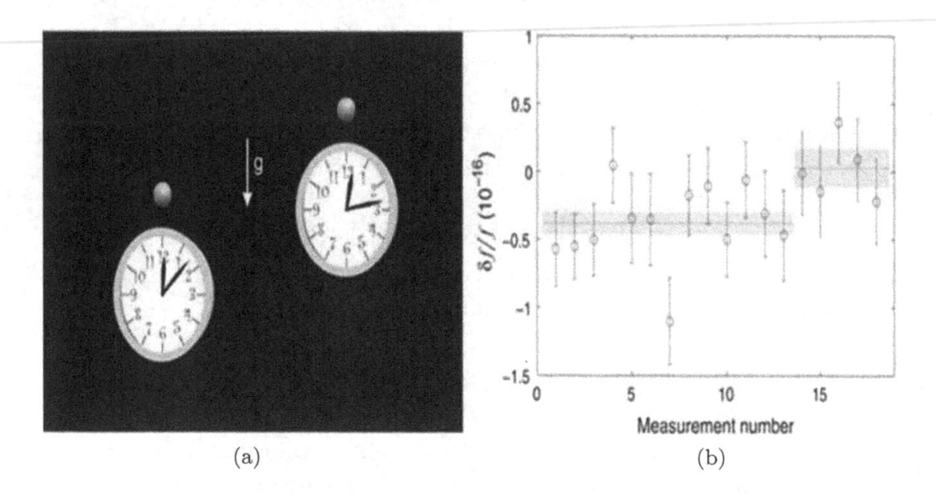

Figure 3.11: Gravitational time dilation at the scale of daily life.

frequency between the two optical clocks of Al^+ at different heights. Net relative change due to height increase is measured to be $(4.1 \pm 1.6) \times 10^{-17}$.

Recently, related optical clocks have achieved a short time stability of about 10^{-18} s/s, cf. Brewer *et al.* (2019).

3.6 The GPS Navigation System

What is relativity good for? It is very common to think that relativity is a rather mysterious mathematical theory that has no consequences for everyday life. This view, in fact, is far from the truth. The Global Position System (GPS) was originally developed by the Department of Defense to provide satellite navigation of the US military system. Later, the GPS was used to establish military control and nowadays exists in many civilian applications in navigation.

The current GPS configuration consists of a network of 24 satellites in high orbits around the Earth. Each satellite in the GPS constellation orbits at an altitude of about 20 000 km above the ground and has an orbital velocity of about 14 000 km/h, hence the orbital period is approximately 12 h. Contrary to popular belief, GPS satellites are not in geosynchronous[2] or geostationary[3] orbits (see Fig. 3.12).

The position determination $(t, \vec{r})$ ("radar coordinates") with four synchronized atomic clocks in satellites involves the principle of the constancy of the velocity c of light via

$$c^2(t - t_i)^2 = |\vec{r} - \vec{r}_c|^2.$$

Satellite orbits are distributed so that at least four satellites are always visible from anywhere on Earth at any given time (with a maximum of 12 visible at the same time). Each satellite carries a quartz oscillator, with a short-term stability better than Caesium clocks. A GPS receiver on an aircraft determines its current position comparing the departure time of the signals it receives from several of the GPS satellites (usually from 6 to 12) and the triangulation of the known positions of each satellite. Accuracy is phenomenal: Even a simple hand-held GPS receiver can determine its

[2]A geosynchronous orbit is a geocentric orbit that has the same orbital period as the Earth's sidereal rotation period. It has a semi-major axis of 42 164 km in the equatorial plane.

[3]A geostationary orbit or GEO is a geosynchronous orbit directly above the Earth's equator, with zero eccentricity. From Earth, a geostationary object seems motionless in the sky and is therefore of interest for artificial satellite operations.

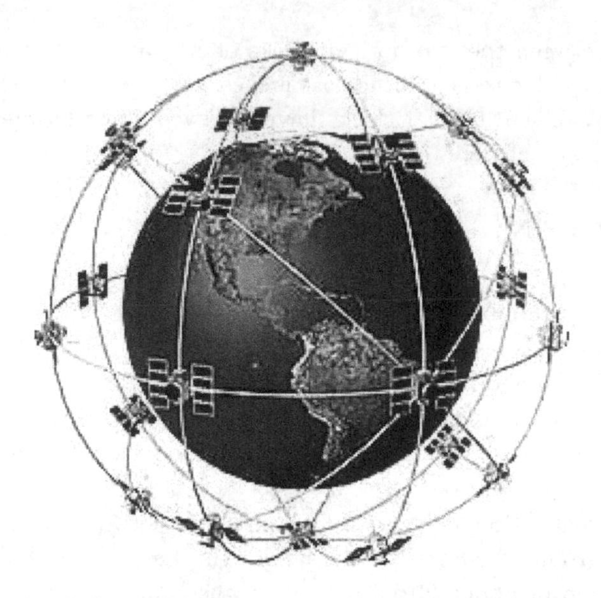

Figure 3.12: The GPS configuration consists of a network of 24 to 32 synchronized satellites that are in orbit around the Earth.

absolute position on the Earth's surface within 5–10 m in just a few seconds. A GPS receiver in a car can provide rather accurate readings of position, velocity, and real-time departure!

More sophisticated is the differential GPS, which delivers position with an accuracy on the level of centimeters.

Absolute and relative sensors, namely a camera, GPS, and heading gyroscope, may guide an autonomous tractor. The system uses a rule-based method to process the sensor data and steer the machine. The system is able to drive the tractor at velocities of 3.6 m/s with less than 15 cm of deviation (Stentz *et al.*, 2002). Near-infrared spectroscopy may facilitate soil monitoring and a more sustainable fertilization.

Also, the mammalian brain uses a sophisticated GPS-like tracking system of its own, allowing animals to learn to navigate and find their way.

There are neuronal cells that allow for spatial navigation, a system arranged in a hexagonal grid. The information from such grid cells, from cells recognizing the direction of the head and the borders of the room in the entorhinal cortex, and from place cells in the hippocampus, form a comprehensive circuitry for a positioning system, a kind of inner GPS in the brain (Moser & Moser, 2016).

3.7 Homework

1. Draw the Lorentz factor γ as a function of the relative velocity $\beta = v/c$.
2. A formula one driver is at the end of a race which took an hour. Assume that the pilot drove at 280 km/h. How long is the pilot younger compared to spectators in the stands?
3. Reproduce the predictions of Hafele and Keating (summarized in the following table), knowing that the planes flying to the east were cruising for 41.2 h at an average height of 8,900 m, while those flying west for 48.6 h had an average height of 9,400 m. Suppose, for simplification, that flights were equatorial with average velocities relative to the ground of 713 km/h to the east and 440 km/h to the west.

Time difference	To the east	To the west
Cinematic dilation	-184 ± 18 ns	96 ± 10 ns
Gravitational dilation	144 ± 14 ns	179 ± 18 ns
Observed effect	-40 ± 23 ns	273 ± 7 ns
Total effect	-59 ± 10 ns	275 ± 21 ns

(A more accurate calculation requires knowing maps of the routes and velocities of the aircraft in different sections in which the flights were subdivided, and contributions from the Sagnac effect, see Ashby (2002).
4. Muons produced by inelastic collisions of cosmic rays with the atoms of the atmosphere have a half-life of 2.3 μs. To travel a $v\Delta\tau_{\rm R} = 4.8$ km distance in the atmosphere, the velocity must be almost that of light, if $\Delta\tau_{\rm R} = 16\mu$ s. Calculate the velocity of the muons (Rossi & Hall, 1941).

References

Brewer *et al.* (2019). "$^{27}Al^+$ Quantum-logic clock with a systematic uncertainty below 10^{-18}", *Phys. Rev. Lett.* **123**, 03320.

Moser, M.-B. and Moser, E.I. (2016). "The brain's GPS tells you where you are and where you've come from", *Sci. Am.*, January.

Stentz, A. *et al.* (2002). "A system for semi-autonomous tractor operations", *Auton. Robots* **13**, 87.

Chapter 4

Relativistic Shape of Mechanics

In this chapter, relativistic concepts are introduced intuitively, this means, without the usage of tensor algebra or 4-vectors, as they will be partially adopted in some of the next chapters.

4.1 Relativistic Momentum and Force

It will be shown later on that the relativistic equation for a particle's linear momentum is

$$\vec{\mathbf{p}} = \frac{m\vec{\mathbf{v}}}{\sqrt{1-(v/c)^2}} = \gamma m\vec{\mathbf{v}},$$

where $\vec{\mathbf{v}}$ is the velocity of the particle, m represents its invariant mass, and $\gamma = 1/\sqrt{1-\vec{\mathbf{v}}\cdot\vec{\mathbf{v}}/c^2}$ is the Lorentz factor. We denote the absolute value of $\vec{\mathbf{v}}$ as: $v \equiv \sqrt{\vec{\mathbf{v}}\cdot\vec{\mathbf{v}}} = |\vec{\mathbf{v}}|$.

Subject to the following physical conditions: The linear momentum of an isolated system is conserved in all collisions. The relativistic linear momentum $\vec{\mathbf{p}}$ of a particle approaches the non-relativistic linear momentum $m\vec{\mathbf{v}}$ when $\mid \vec{\mathbf{v}} \mid \ll c$.

The relativistic force $\vec{\mathbf{F}}$ which acts on a particle of constant mass m is defined as

$$\begin{aligned}
\vec{\mathbf{F}} \equiv \frac{d\vec{\mathbf{p}}}{dt} &= \frac{d}{dt}\left(\frac{m\vec{\mathbf{v}}}{\sqrt{1-\vec{\mathbf{v}}\cdot\vec{\mathbf{v}}/c^2}}\right) \\
&= m\left[\frac{\dot{\vec{\mathbf{v}}}}{\sqrt{1-\vec{\mathbf{v}}\cdot\vec{\mathbf{v}}/c^2}} - \frac{1}{2}\frac{\vec{\mathbf{v}}(-2\vec{\mathbf{v}}\cdot\dot{\vec{\mathbf{v}}}/c^2)}{(1-\vec{\mathbf{v}}\cdot\vec{\mathbf{v}}/c^2)^{3/2}}\right].
\end{aligned}$$

Here the dot over the vectors indicates a time derivative, i.e. $\cdot \equiv \partial/\partial t$. Equivalently, applying Leibniz's rule, we get

$$\vec{\mathbf{F}} \equiv m\left[\gamma\frac{d\vec{\mathbf{v}}}{dt} + \vec{\mathbf{v}}\frac{d\gamma}{dt}\right].$$

Its relation to the Minkowski force will be exhibited in a separate chapter. For accelerated uniform rectilinear motion, i.e. where $\vec{\mathbf{v}}\cdot\vec{\mathbf{a}} = |\vec{\mathbf{v}}||\vec{\mathbf{a}}|$ holds, the absolute value of the force reduces to

$$F \equiv \left|\vec{\mathbf{F}}\right| = \frac{m\,|\vec{\mathbf{a}}|}{(1-\vec{\mathbf{v}}\cdot\vec{\mathbf{v}}/c^2)^{3/2}} \simeq m\,|\vec{\mathbf{a}}|.$$

By applying a *constant* force $\vec{\mathbf{F}}$ to a massive particle, its acceleration $\vec{\mathbf{a}} \equiv d\vec{\mathbf{v}}/dt$ decreases, in which case, $|\vec{\mathbf{a}}| \propto \left(1-\vec{\mathbf{v}}\cdot\vec{\mathbf{v}}/c^2\right)^{3/2}$. Given this proportionality relation, note that when the velocity of a particle approaches that of the light c, the acceleration caused by any finite force approaches zero. Consequently, it would be impossible to accelerate a massive particle initially at rest to a velocity $v \geq c$ exceeding the velocity of light.

4.2 Relativistic Energy

The work done by a force $\vec{\mathbf{F}}$ on a particle is equal to the integral

$$K = \int_{x_1}^{x_2} F dx = \int_{x_1}^{x_2} \frac{dp}{dt}\, dx$$

along, e.g. the x direction. Replacing the change rate $dp/dt = F$ by the above norm and $dx = vdt$ we get

$$K := \int_0^t \frac{m(dv/dt)vdt}{(1-\vec{\mathbf{v}}\cdot\vec{\mathbf{v}}/c^2)^{3/2}} = m\int_0^v \frac{vdv}{(1-\vec{\mathbf{v}}\cdot\vec{\mathbf{v}}/c^2)^{3/2}}$$
$$= \frac{mc^2}{\sqrt{1-\vec{\mathbf{v}}\cdot\vec{\mathbf{v}}/c^2}} - mc^2.$$

Suppose that the particle is accelerated from rest to some final velocity $v < c$, then the relativistic kinetic energy[1] is

$$K = \frac{mc^2}{\sqrt{1-\vec{\mathbf{v}}\cdot\vec{\mathbf{v}}/c^2}} - mc^2 = \gamma mc^2 - mc^2 = (\gamma - 1)\, mc^2.$$

[1]Since $K = c\left[\sqrt{p_0 p^0} - \sqrt{p_\mu p^\mu}\right]$, the kinetic energy is not invariant under a Lorentz transformation, only $mc = \sqrt{p_\mu p^\mu}$ is.

By using a binomial expansion[2] it can be shown that at small velocities, i.e. when $v/c \ll 1$, this equation reduces to the classical non-relativistic Newton expression $K_{\rm N} = \frac{1}{2}mv^2$.

In this case, the Lorentz factor can be expanded as follows (for $\beta = v/c \ll 1$):

$$\gamma = \frac{1}{\sqrt{1-\frac{\vec{v}\cdot\vec{v}}{c^2}}} = \left(1 - \frac{\vec{v}\cdot\vec{v}}{c^2}\right)^{-1/2} \approx 1 + \frac{1}{2}\frac{\vec{v}\cdot\vec{v}}{c^2}.$$

By substituting this in the equation for the kinetic energy, we get

$$K \approx \left[\left(1 + \frac{1}{2}\frac{\vec{v}\cdot\vec{v}}{c^2}\right) - 1\right] mc^2 \simeq \frac{1}{2}mv^2,$$

which is the classical expression for the kinetic energy. In Fig. 4.1, relativistic and non-relativistic expressions are compared. According to a

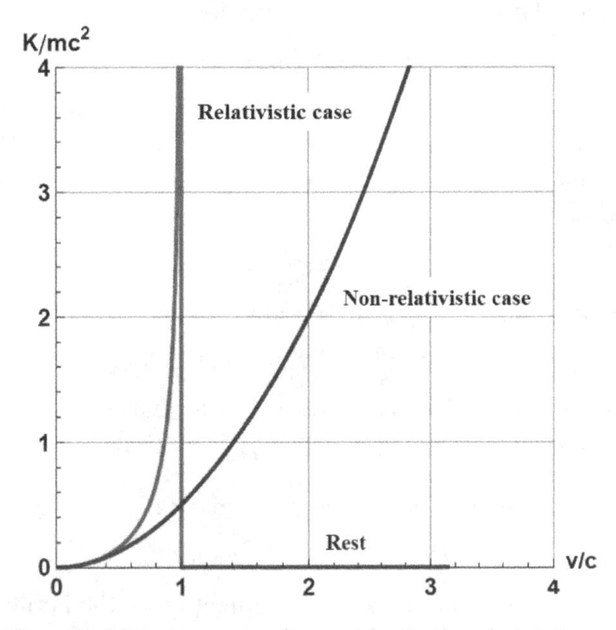

Figure 4.1: Normalized kinetic energy of a particle. In the relativistic case, v is always less than c, i.e. $\beta = v/c \leq 1$.

[2]The binomial expansion $(1-\beta^2)^{-1/2} \approx 1 + \frac{1}{2}\beta^2 + \cdots$ can be used when $\beta \ll 1$, and higher order terms for β can be neglected. In relativistic mechanics, β is an adimensional variable which represents the quotient v/c.

relativistic approach, the velocity of a particle never exceeds the velocity of the light c, regardless of its kinetic energy. Both curves align if $v \ll c$.

Since the total energy is equal to the kinetic energy plus the energy at rest of a particle, we can define the energy[3] E as

$$E \equiv K + mc^2 = \gamma mc^2 = \frac{mc^2}{\sqrt{1 - \vec{\mathbf{v}} \cdot \vec{\mathbf{v}}/c^2}}.$$

Squaring E,

$$E^2 = \vec{\mathbf{p}} \cdot \vec{\mathbf{p}} c^2 + (mc^2)^2$$

results. This formula corresponds to the Pythagorean theorem in the "relativistic triangle".

4.2.1 *Unitary transformation in QED*

In quantum electrodynamics (QEDs), this triangle also occurs in an approximation of the relativistic (free) Dirac equation

$$i\gamma^\mu \partial_\mu \psi = (mc/\hbar)\psi.$$

After applying the Foldy–Wouthuysen (FW) unitary tranformation (Hehl *et al.*, 1991) and the parametrization $\tan 2\theta = |\vec{\mathbf{p}}|/mc$, the Hamiltonian in Dirac's equation can be diagonalized leading to the non-relativistic approximation

$$\hat{H}' \simeq \gamma^0 E,$$

where γ^0 is the temporal Dirac matrix entering in the Clifford algebra:

$$\gamma^\mu \gamma^\nu + \gamma^\nu \gamma^\mu = 2\eta^{\mu\nu}.$$

The FW transformation is a unitary transformation

$$\psi \longrightarrow \psi' = U\psi$$

of the Fermion wave function via the 4×4 matrix

$$U = \mathbb{1}_4 \cos\theta + \vec{\gamma} \cdot \hat{p} \sin\theta.$$

Here $\hat{p} = \vec{p}/|\vec{p}|$ is a unit vector in the direction of the Fermion momentum. Applying this unitary transformation to the free Dirac Hamiltonian,

$$\hat{H} = c\gamma^0(\vec{\gamma} \cdot \vec{p} + mc),$$

[3]For tachyons (hypothetical particles) which travel faster than the velocity of light, their mass $\mu \equiv im$ is purely imaginary and their kinetic energy $E = \mu c^2/\sqrt{\beta^2 - 1}$ would approach zero as the velocity goes to infinity, i.e. $\beta \to \infty$.

we obtain:

$$\hat{H}' = U\hat{H}U^{-1} = c\gamma^0(\vec{\gamma}\cdot\vec{p} + mc)(\cos 2\theta - \gamma^0\vec{\gamma}\cdot\hat{p}\sin 2\theta).$$

In view of the Clifford algebra, this factors out into

$$\hat{H}' = c\gamma^0\vec{\gamma}\cdot\vec{p}\left(\cos 2\theta - \frac{mc}{|\vec{p}|}\sin 2\theta\right) + c\gamma^0(mc\cos 2\theta + |\vec{p}|\sin 2\theta).$$

This unitary transformation is continuous in θ. Let us *diagonalize* the transformed Hamiltonian by choosing

$$\tan 2\theta = \frac{|\vec{p}|}{mc},$$

as in the relativistic triangle in Fig. 4.2. Then, by elementary trigonometry, using $\cos 2\theta = mc^2/E$ and $\sin 2\theta = pc/E$ of the relativistic triangle again,

$$\hat{H}' = c\gamma^0(mc\cos 2\theta + |\vec{p}|\sin 2\theta) = c\gamma^0\sqrt{\vec{p}^2 + m^2c^2}$$

results. This was known before as the Newton–Wigner representation of Dirac's equation.

When the particle is at rest, i.e. $\vec{\mathbf{p}} = 0$, one gets

$$E_0 = mc^2.$$

This is referred to as the rest energy and shows that mass is a form of energy,[4] at times called "mass and energy equivalence". The parameter m is known as invariant mass since it remains unchanged under a Lorentz transformation.

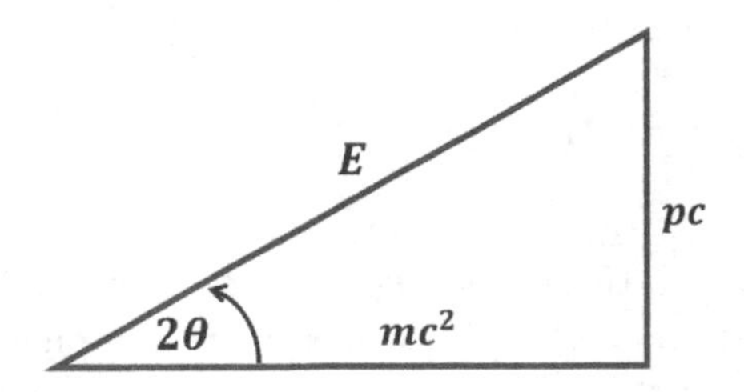

Figure 4.2: The "relativistic triangle", as a mnemotechnique in order to remember the famous formula of Einstein.

[4]The popular equation "$E = mc^2$" for the total energy is false and was never used by Einstein, see Okun (1989), for a more elaborated explanation of the concepts regarding mass. Group-theoretically, p^2 is an (invariant) Casimir operator of the inhomogeneous Lorentz group which, by Schur's lemma, is a constant, i.e. m^2c^2 in our case.

For massless particles, such as photons, we put $m = 0$, cf. Goldhaber & Nieto (2010) and find that

$$E = |\vec{\mathbf{p}}|\, c.$$

4.2.2 *Examples*

4.2.2.1 *Rest energy of an electron*

The invariant mass of an electron is $m_e = 9.11 \times 10^{-31}$kg. Therefore, its energy at rest is, approximately:

$$\begin{aligned} E_0 &= m_e c^2 = (9.11 \times 10^{-31}\text{kg})(3.00 \times 10^{18}\text{m/s})^2 = 8.20 \times 10^{-14}\,\text{J} \\ &= (8.20 \times 10^{-14}\text{J})(1\,\text{eV}/1.60 \times 10^{-19}\text{J}) = 0.511\,\text{MeV}, \end{aligned}$$

where $1\,\text{eV} = 1.60 \times 10^{-19}$ J is the conversion factor.

4.2.2.2 *Fast moving electron*

An electron inside a cathode ray tube or inside an electron microscope shall move with an approximate velocity of $v = 0.250c$. Find its kinetic energy and its total energy in eV.

Solution: Using the electron's rest energy $E_0 = 0.511\,\text{MeV}$ along with the equation previously derived, we have

$$E = \frac{m_e c^2}{\sqrt{1 - \frac{v^2}{c^2}}} = \frac{0.511\,\text{MeV}}{\sqrt{1 - \frac{(0.250c)^2}{c^2}}} = 1.03(0.511\,\text{MeV}) = 0.528\,\text{MeV}.$$

This is 3% higher than its rest energy.

4.3 Homework

1. (a) Find the rest energy of a proton in electron-Volts (eV). (b) If the total energy of a proton is three times its rest energy, what is the velocity of the proton? (c) Determine the kinetic energy of a proton in eV. (d) Compute the momentum of this proton.
2. Compute the momentum of an electron moving with a velocity of (a) 0.010c, (b) 0.500c, and of (c) 0.900c.
3. The non-relativistic expression for the momentum of a particle $\vec{\mathbf{p}} \cong m\vec{\mathbf{v}}$ agrees with experiments only if $v \ll c$. For which velocity does this equation produce an error in the relativistic momentum of (a) 1.0% and of (b) 10%?

4. A golf ball moves with a velocity of 90.0 m/s. By what fraction does its relativistic momentum p differ from its classical value mv? In other words, find the ratio $(p - mv)/mv$.
5. Show that the absolute value of the velocity of an object with a moment p and mass m is $v = c/\sqrt{1 + (mc/p)^2}$, cf. Rindler (1991).
6. Using the expression $\vec{\mathbf{F}} \equiv m\left[\gamma \frac{d\vec{\mathbf{v}}}{dt} + \vec{\mathbf{v}} \frac{d\gamma}{dt}\right]$, show that the force in three dimensions is not always proportional to the acceleration, namely, $\vec{\mathbf{F}} = m\gamma\left[\vec{\mathbf{a}} + \gamma^2 \frac{\vec{\mathbf{v}} \cdot \vec{\mathbf{a}}}{c^2} \vec{\mathbf{v}}\right]$. However, $\vec{\mathbf{F}} \cdot \vec{\mathbf{v}} = m\gamma^3 \vec{\mathbf{a}} \cdot \vec{\mathbf{v}}$ holds for the scalar product.

Reference

Hehl, F.W. *et al.* (1991). "Two lectures on fermions and gravity", in: *Geometry and Theoretical Physics*, Bad Honnef Lectures, 12–16 February 1990, J. Debrus and A.C. Hirshfeld eds. (Springer, Berlin, 1991), pp. 56–140.

Chapter 5

Spacetime Diagrams

Einstein's work, *On the Electrodynamics of Moving Bodies*, written in 1905, presented two postulates that comprise the foundation of his Special theory of Relativity (SR).[1]

- **Principle of relativity**

  The laws of physics are the same in all inertial reference frames.[2]

Originally, this postulate refers to the impossibility of distinguishing between reference frames at rest or in constant motion. There is no way to know the state of motion of an observer from any physical experiment that is performed by that observer within his/her reference system (if we play a football game on a boat or a plane in uniform motion, not accelerated, it is the same as if we played it on the ground. The laws of physics do not change). Therefore, an absolute reference system does not exist in SR, inasmuch as each object with uniform motion can be used as a reference system for the "rest of the universe" without changing the laws of physics at all.

As an example of the invariance of the laws of physics, the electron charge is the same in different laboratories on Earth (disregarding her acceleration). This is a part of the $\mathbb{CPT}$ theorem in relativistic quantum field theory (QFT).

[1] It is called Special Relativity as it applies only to non-accelerated observers. The General Relativity, which Einstein developed later, also includes accelerated frames. Unlike the first, which was published in a single paper in Annalen der Physik in 1905, GR was established over several years between 1907 and 1915.

[2] Ludwig Lange was the first in conceptualizing inertial frames, see Von Laue (1948).

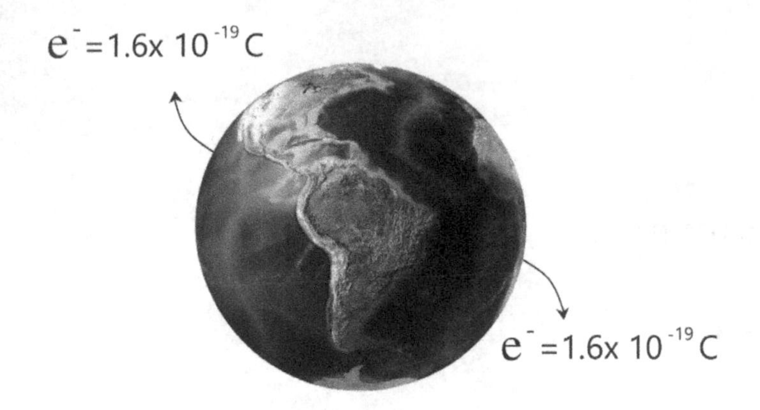

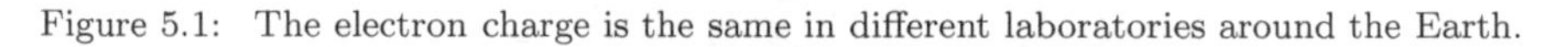

Figure 5.1: The electron charge is the same in different laboratories around the Earth.

This principle refers not only to laws describing moving objects (for them it is easy to demonstrate), but to all the laws of physics, which makes it a postulate (not demonstrable) that has not been experimentally refuted yet. With this postulate, Einstein put an end to the idea of the luminiferous æther, that was intended to constitute an absolute frame of reference. If it existed, Maxwell's equations (which predict the velocity of light and do not need any specification of the observer's velocity) would depend on the observer's state of motion relative to an æther.

- **Principle of universality of the velocity of light**

  The velocity of light in a vacuum has the same value c in all inertial frames.

The recognized success of Maxwell's equations in describing a wide variety of electromagnetic phenomena[3] and the consistency of this principle with the behavior of fermionic matter described by the relativistic Dirac equation, suggested the validity of this postulate.

However, as we have already seen, the first direct experimental confirmation of this postulate did not arrive until 1964 when it was realized that photons emitted by a high-velocity pion did not travel at different velocities than those emitted by a radioactive source at rest. This is responsible for

[3]The French mathematician d'Alembert considered already in 1747 time as a "fourth dimension" and Bernhard Riemann (1867), based on his wave equation, a preliminary scalar form of electrodynamics.

making it difficult to conciliate the theory of relativity with our vision of the world through our senses, evolutionarily developed for small velocities.

5.1 Spacetime: Minkowski Diagrams

The unified concept of spacetime, introduced by Hermann Minkowski[4] in 1908, is a mathematical idealization. Previously, space and time were considered as separate, because both are measured in very different ways and perceived differently as well.

However, in Special Relativity, in order to specify an event one has to say where (three spatial dimensions) and when (one more dimension, time). Minkowski proposed to conceive the world as a four-dimensional (4D) spacetime network. This view has two advantages: First, it leads us to a simple and practical graphical resolution of Lorentz transformations, by using spacetime or *Minkowski diagrams*. In addition, these spacetime diagrams allow us to visualize the evolution of an object in space and time as a "complete film" via its succession line. Considering for illustrative purposes a velocity of light of $c = 1$ meter per second, the spacetime diagram for a photon is as follows:

On the same diagram, specifying the $\vec{x}$ position of a particle and the time t for its position thus is representing an event in relativity (that is, a point in the plane represents an event).

Let us reduce, for illustrative purposes, the three spatial coordinates $\vec{x} = (x, 0, 0)$ to one. Then the set of events in the plane (x, t) that represents a particle at several moments is known in relativity as a worldline. In SR, the line of events is always a straight line (such as the red line shown in Fig. 5.2) if the material particle travels with constant velocity $v = dx/dt$, thus traveling equal distances at equal times.

Usually a spacetime diagram is depicted in 2D and both axes have the same type of units, that is, both the vertical and horizontal axes have units of length. It is a convention to multiply the time axis by the absolute universal constant c which is the velocity of light. Then ct becomes a distance measured in light-seconds (Ls). In all spacetime diagrams to be used here, following Feynman, the vertical axis will be the time axis, in physical dimensions of meters.

[4] H. Minkowski was a German mathematician who created the geometrical theory of numbers. Contributing to number theory, mathematical physics and theory of relativity, thereby he provided the mathematical foundations for the Special Theory of Relativity of his former student A. Einstein.

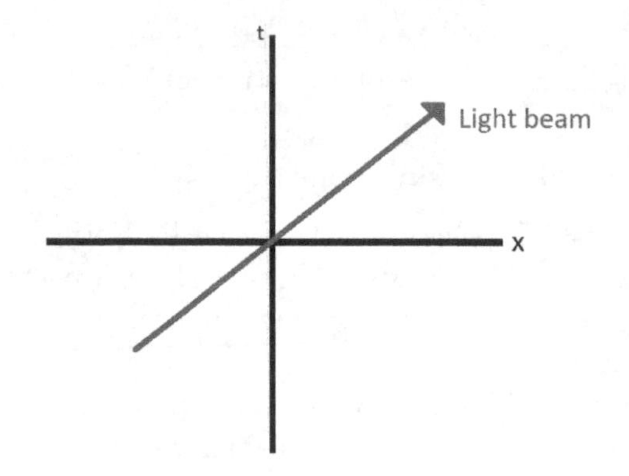

Figure 5.2: Spacetime diagram for a light beam.

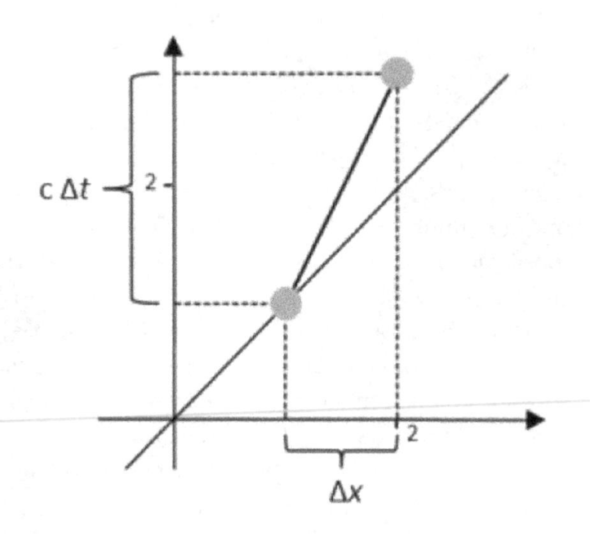

Figure 5.3: Graph of two distinct events that are represented by gray dots.

In Fig. 5.3, we have represented two different events, one occurring at the position x_1 at a time t_1 and the other event occurring at the position x_2 at another time t_2. By setting numbers and using a velocity of light equal to $c = 1$ meter/second, the respective coordinates of each event and the distance between both events will have "human" size:

$$\Delta x = x_2 - x_1 = 2\,\mathrm{m} - 1\,\mathrm{m} = 1\,\mathrm{m},$$

$$c\Delta t = ct_2 - ct_1 = 3\,\mathrm{m} - 1\,\mathrm{m} = 2\,\mathrm{m}.$$

5.1.1 *Observer at rest*

An inertial observer $\mathcal{O}$ is any non-accelerated observer. According to the principle of relativity, there is no privileged observer. Please note that *an observer at rest is represented by a specific inertial frame.*

We locate an *event* A using a point whose spatial coordinate x, and temporal, ct, can be read on the coordinate axes of the spacetime diagram. The ct coordinate indicates the event's coordinate time, and x is the distance measured from the origin that is taken as a reference point. The x-axis is the set of simultaneous events that occur at $ct = 0$. Any parallel line to the x-axis ($ct = ct_0$) indicates simultaneous events that occur at another instant of time t_0.

World lines of photons are represented by lines inclined by 45°, because for them $ct = x$ or $ct = -x$, depending on the direction of light propagation (i.e., from left to right or from right to left, respectively), when using dimensionless coordinates with $c = 1$. The *line of events* for an object that moves at uniform velocity v is a straight line of $x = vt$ or $ct = \dfrac{c}{v}x$ forming an angle $\phi = \arctan(x/ct) = \arctan(v/c)$ with the ct-axis. The sign is positive

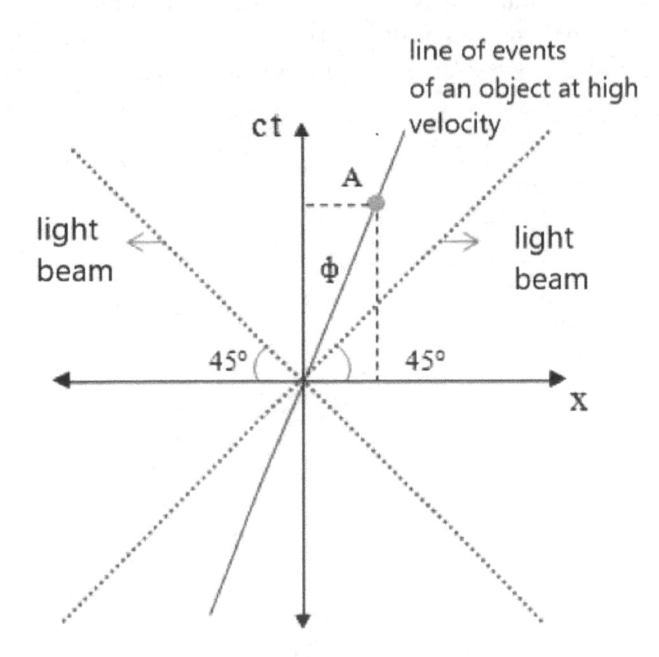

Figure 5.4: A represents an event, while the red line represents the line of events of an object at high velocity but less than that of light.

or negative whether you move from left to right or right to left, respectively. We will see that the angle ϕ in absolute values is always $|\phi| \leq \pi/4$. If the object's world line is not straight, then the motion is not uniform. For a photon (massless particle) with $v = c$, we get $\phi = \pm\pi/4$.

5.1.2 *Observer in relative motion: Lorentz transformations*

So far we have described particles according to an observer at rest, for example, with regard to the tracks of a train. Let us see how to draw the spacetime diagram for another observer moving uniformly in a train at velocity v, relative to an observer at rest. For simplicity, we match the origin of coordinates of both observers.

The x'-axis is the set of simultaneous events that occur at $ct' = 0$, which is the same as the $t = vx$ line. Therefore, it forms an angle $\phi = \arctan v/c$ with the x-axis.

The axis ct' is the set of events that occur at the place $x' = 0$, which is the same as the $t = x/v$ line. Therefore, it forms the same angle $\phi = \arctan v/c$, this time with the ct axis.

The spatiotemporal coordinates of an event, for example, the event A that we studied before, are parallel to the x' and ct' axes, which now shape a diamond (Fig. 5.5) whose lines will not anymore be perpendicular to each other.

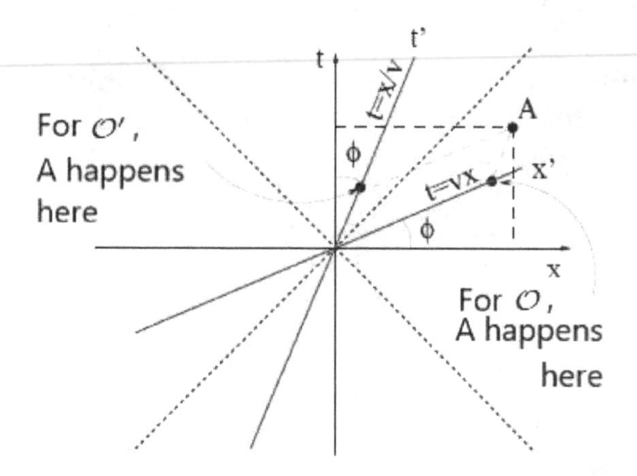

Figure 5.5: Spacetime diagram for an observer $\mathcal{O}'$ in motion whose reference system is (ct', x'). The coordinates for the same event A from two different reference frames have been drawn parallel to their respective axes.

5.2 Spacetime Interval

Apparently, it is all "relative for the observer". However, we have already seen that the velocity of light is the same for any observer. Based on this, we have another important invariant: This is the quadratic interval between two events, which any observer can easily determine (via radar signals) from measurements of locations in space and time. Suppose, for simplicity, that one of the two events is the spatiotemporal origin $\mathcal{O}$, which coincides with $\mathcal{O}'$. Let (x, ct) and (x', ct') be the coordinates of another event A, depending on each reference system or observer. The invariant interval squared is

$$\text{Interval squared} \equiv (ct)^2 - x^2 = (ct')^2 - x'^2.$$

The Lorentz transformations in 2D that connect the two inertial frames must satisfy the equations

$$t' = \gamma\left(t - \frac{vx}{c^2}\right),$$
$$x' = \gamma(x - vt),$$

where

$$\gamma = 1/\sqrt{1 - v^2/c^2}$$

is the Lorentz factor.

Since these transformations leave, in 2D, the quadratic interval unchanged, SR is really a theory of spacetime invariants!

5.2.1 *The calibration of the axes*

The spacetime interval helps us to calibrate the axes: The distances between the marks of the axes for each observer do not measure the same unit, i.e. a second in the reference system $\mathcal{O}$ is different from one second in the system of $\mathcal{O}'$. As a unit, we will use the light-second (Ls).

To find the relationship between the "tics" on the time axis, just look at the points where the hyperbola $(ct)^2 - x^2 = (\text{Ls})^2$ cut the ct' axis, given by $ct = vx/c$, as ($x = 0$, $ct = \text{Ls}$) transforms into ($x' = 0$, $ct' = \text{Ls}$).

To continue with the calibration of the time axis, we only have to take the intersections of the hyperbolas

$$(ct)^2 - x^2 = (n\text{Ls})^2, \quad n = 1, 2, 3, \ldots,$$

with the temporal axis ct' given by $t = x/v$, because the spacetime event ($x = 0$, $ct = n\text{Ls}$) transforms into ($x' = 0$, $ct' = n\text{Ls}$). In the same manner,

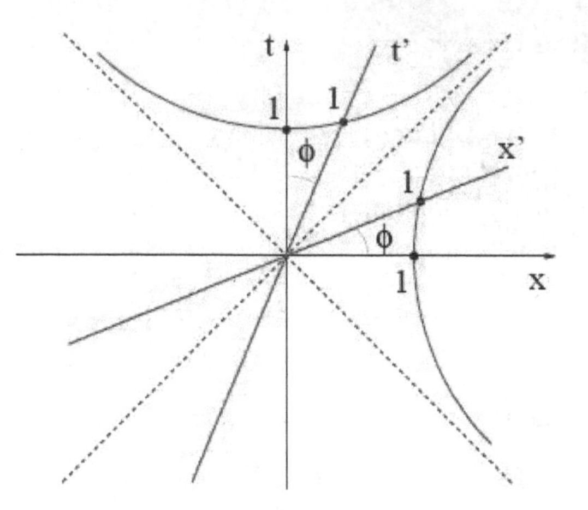

Figure 5.6: Axes calibrations for the observer $\mathcal{O}'$ for any (ct', x') system.

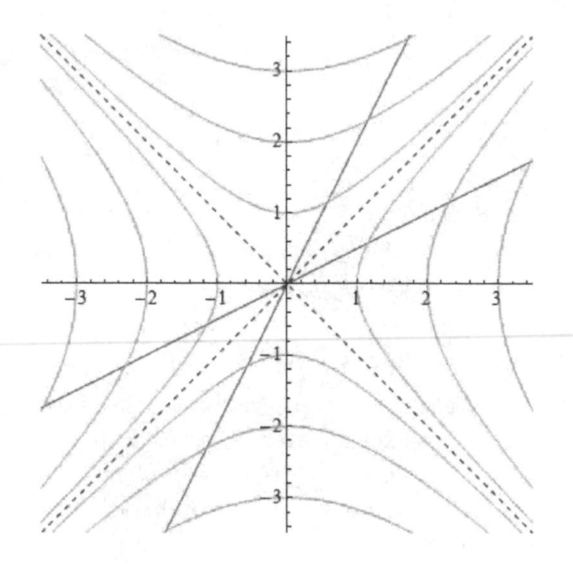

Figure 5.7: Calibrated axes across the Cartesian plane for observers $\mathcal{O}$ and $\mathcal{O}'$.

take the intersections of the hyperbolas.

$$(ct)^2 - x^2 = -(n\mathrm{Ls})^2 \quad n = 1, 2, 3, \ldots,$$

with the spatial axis x', given by $ct = vx/c$, because ($x = n\mathrm{Ls}$, $ct = 0$) transforms into ($x' = n\mathrm{Ls}$, $ct' = 0$). Where $n\,\mathrm{Ls}$ (multiples of light-second) are the intervals we need for a calibration (Liebscher, 1977).

5.2.2 *Example: Space rocket*

A spaceship of the fictitious length of one light second (1Ls) travels with the velocity of $v = 3c/5$ relative to the laboratory system I. The axes of the frame I′ can be build as follows:

(1) Using the transformations of Lorentz in 2D, we can obtain the equations that will allow us to build the axes of the I' system. The slope of the line is related to the relative velocity. If v is close to the velocity of light, the t' axis will approach the world line of light that bisects the system I. It rotates away from the lightcone if the relative velocity gets much lower than c. Equations for the axes are:

- Axis t' is the world line for which $x' = 0 \rightarrow x = vt$,
- Axis x' is the world line for which $t' = 0 \rightarrow t = \dfrac{v}{c^2}x$.

Another equivalent representation of the new axes is geometrically determined by their rotation angle:

$$\tan\delta = \frac{v}{c} = \tan\delta'' \quad \text{or} \quad \tan\delta' = \frac{c}{v}.$$

(2) To fix the unit second on the t' axis, we must consider $t' = t/\gamma$, therefore, $t = 1s$ in the I system merely would correspond to

$$t' = 1\text{s}\sqrt{1 - (\vec{v}/c)^2} = 1\text{s}\sqrt{1 - (3/5)^2} = 4/5\text{s}.$$

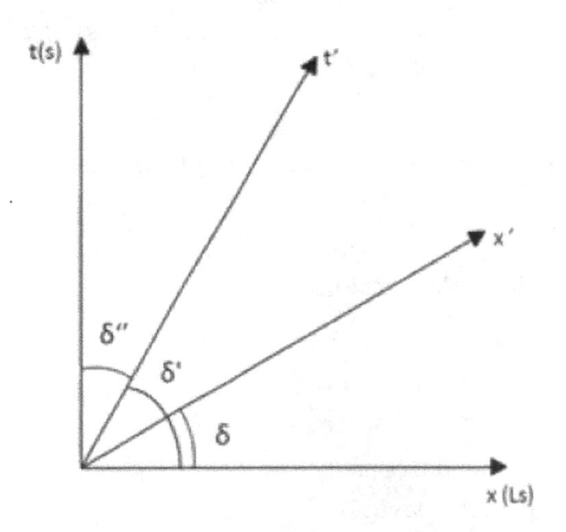

Figure 5.8: Rotation angle of the axes.

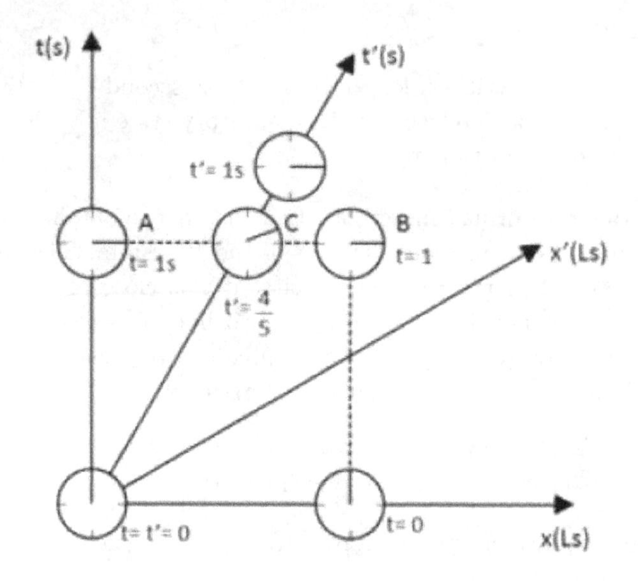

Figure 5.9: To fix $t' = 1$ s, we need to consider the time dilation.

Inversely, in order to obtain $t' = 1\,\text{s}$ in the moving frame, we need to multiply it by $\gamma = 1/\sqrt{1 - (\vec{v}/c)^2}$, getting $t = 5/4\,\text{s}$ due to time dilation.

(3) For the rocket with a fictitious length of 1 Ls, the unit $x' = 1\text{Ls}$ has the same extension as $t' = 1$ s due to the constancy of the velocity of light, therefore, Fig. 5.10 is symmetrical. Because of the congruence of triangles, a *Lorentz contraction*

$$L = L_0\sqrt{1 - \vec{v}^2/c^2}$$

results. Physically, it is a consequence of time dilation and simultaneity.

5.2.3 *Hyperbolic intersections*

All units of inertial frames can be inferred from the following relativistic invariant in 2D:

$$c^2t^2 - x^2 \equiv c^2t'^2 - x'^2.$$

Particularly, if $t' = 1$ and $x' = 0$, the system is at rest. In natural units with $c = 1$, a hyperbolic equation,

$$t^2 - x^2 = 1,$$

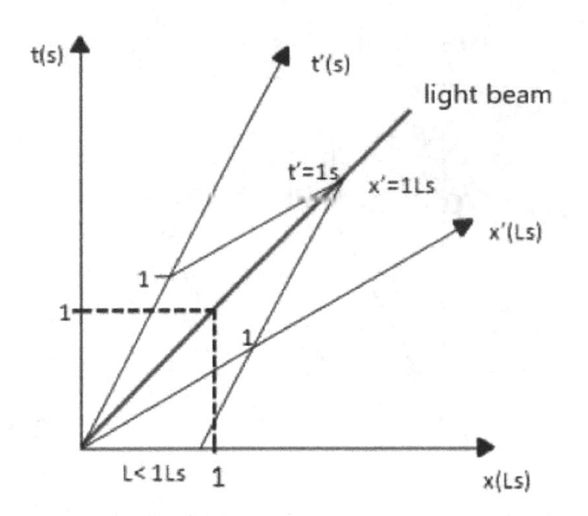

Figure 5.10: The square of units in frame I becomes a rhombus in the frame I', with the light cone as the major bisector.

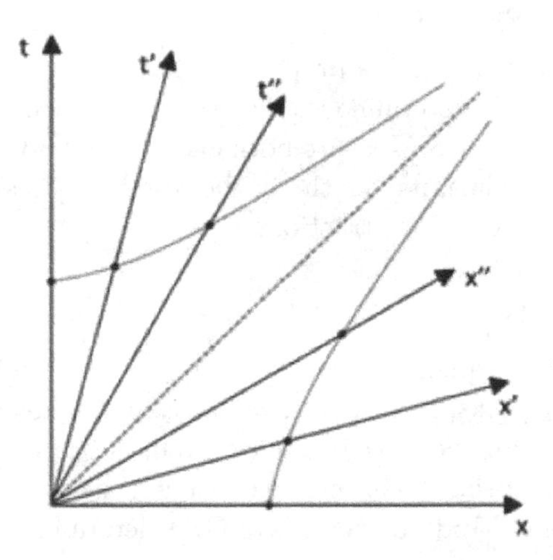

Figure 5.11: Units representation on (x, t).

results that helps us represent the time units in the I' system. In order to construct the spatial units, we take $t' = 0$ and $x' = 1$, yielding the inverted hyperbola as follows:

$$x^2 - t^2 = 1.$$

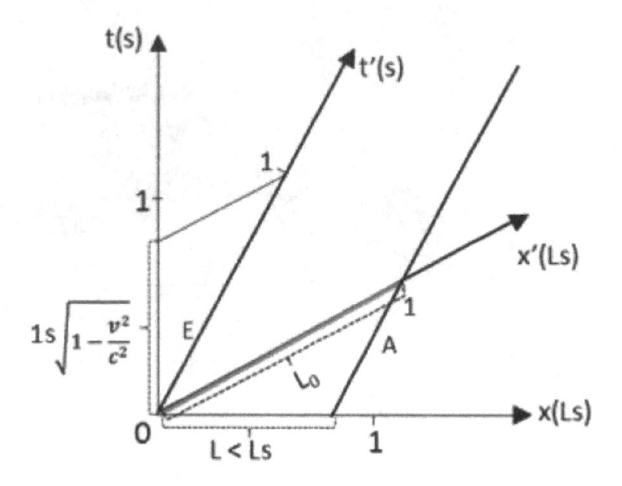

Figure 5.12: In the rest frame, L is smaller than L_0 as a result of the Lorentz–FitzGerald contraction.

5.2.4 *Lorentz contraction*

Example: A spaceship with a proper length $L_0 = 1\,\text{Ls}$ is moving with the velocity $v = 3c/5$. To determine the length of the spaceship (distance $\overline{AE}$) in the rest frame, we must measure both ends of the body at *the same time* via parallels to t'. Thus, its length[5] in the rest frame is smaller ($L < 1\text{Ls}$), as a result of the Lorentz contraction.

5.3 Simultaneity

The concept of simultaneity is the key to understanding relativity (Einstein, 1915): Although clocks, in principle, can be compared by a very slow (adiabatic) transport from one place to another, radio signals are now facilitating an intrinsic comparison. They need the same time for a *halfway distance* between two points. Indeed, nowadays GPS determines places and times with great precision (if engineers do not make mistakes as in the case of "superluminal neutrinos" where a microwave cable was poorly connected).

Construction of the axes in a new inertial frame:

- ***t'*-axis:** The end E of the rocket moving with velocity v with respect to I is at rest relative to I' and must be parallel to the new time axis, so we identify it with t'.

[5] An extended body will appear rotated in 3D due to effects of relativistic optics.

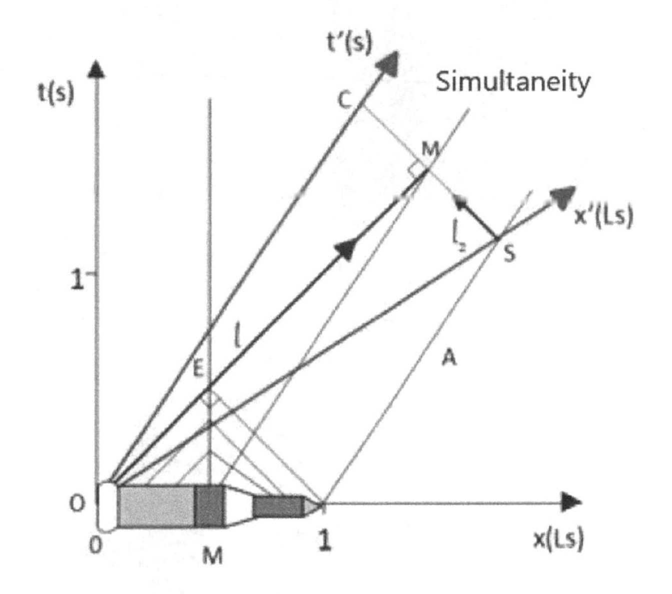

Figure 5.13: A rocket with a length of 1 light-second and velocity $v = 3c/5$.

- **x'-axis:** For the point M located halfway up the rocket, events O and S occur at the same time relative to the lab frame I. Therefore, the axis $x' = \overline{OS}$. The x'-axis is by definition the place of the simultaneous events for time $t' = 0$. Also in I', the light emitted from the origin 0 is a bisector for the new coordinate system (t', x').

The intersections of these coordinates with the past light cone, having its vertex at the midpoint M, are simultaneous events. This standard Einstein simultaneity can be proven to be unique (Giulini, 2001) for inertial frames. To this end, group-theoretical methods are used, as summarized in Appendix A.

5.4 Temporal Order: Past, Present, Future, and Causality

Using spacetime diagrams, it is easy to see that simultaneous events for one observer are not simultaneous for another. For example, events O and C in Fig. 5.14. This is the relativity of the simultaneity.

Now there is something we should be worried about. There are events that follow the same temporal order for two inertial observers while others change order (Fig. 5.14). However, we expect that some events keep

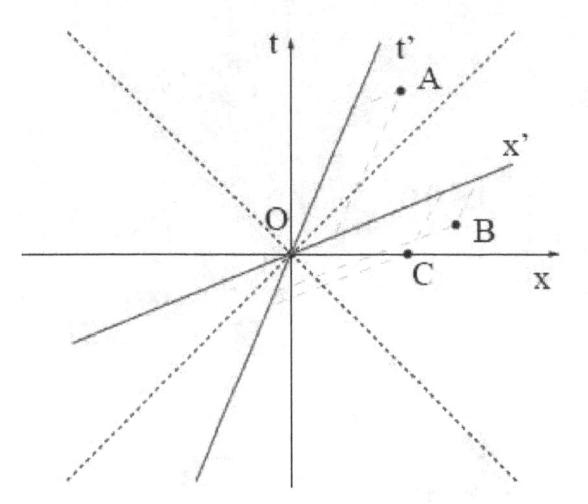

Figure 5.14: Events O and C are simultaneous for $\mathcal{O}$ but not for $\mathcal{O}'$. Event A occurs after O, for both $\mathcal{O}$ and $\mathcal{O}'$. Event B occurs after O for $\mathcal{O}$, but before O for $\mathcal{O}'$.

the temporal order for any inertial observer. We refer to those which are causally related: otherwise we would live in a mystical world in which the effects may precede their causes depending on the relative velocity with which we observe them.

The region of events in spacetime causally connected with an event O at the origin is shown in Fig. 5.15. To demonstrate this, we simply draw the axes of an inertial observer that moves with arbitrary velocity, but never faster than light. It is then easy to see that events located above the lines inclined at 45° always keep the same temporal order: In 3D this is the *light cone* observed at the origin of the coordinate system.

Note that causally connected events are separated by a positive interval (according to our interval definition), which we call time-like. There is no inertial observer who can measure temporarily separated events as simultaneous events.

Non-causally connected events are separated by a negative interval, which we call space-like. It is always possible to find an inertial observer who can measure spatially separated events as simultaneous events. The temporal order of two events depends on the observer.

Events connected by light or radio signals are separated by a null or light-like interval; physically an expanding spherical wave front. Einstein radically changed our concept of the past, present, and future, introducing a new subdivision: For an event O, the past exists (bottom interior of the

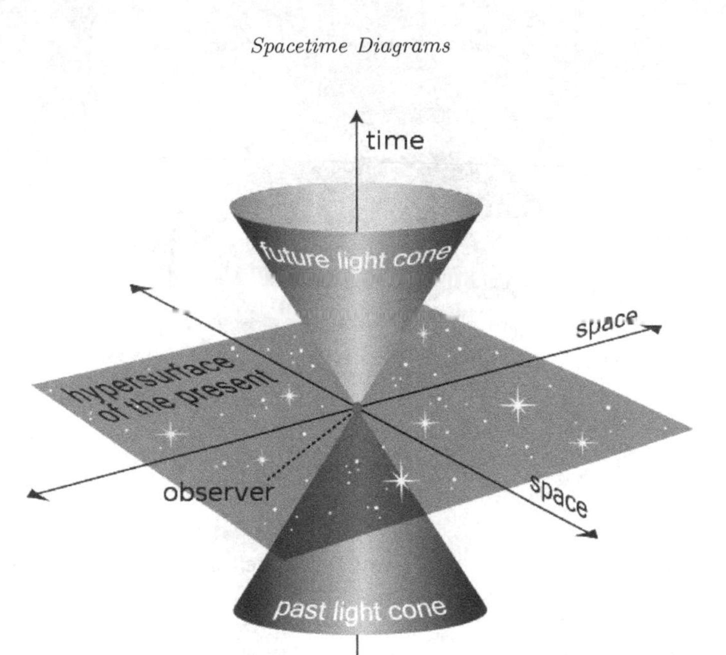

Figure 5.15: Region of events in 3D spacetime causally connected at the origin.

light cone), the present (vertex of the light cone), and the future (part of the top of the light cone). Anything else (outside the light cone) does not make any physical sense! The latter subdivision contains events that cannot affect O and also those that O will never affect.

$$\text{Time-like interval: } (ct)^2 - x^2 = (ct')^2 - x'^2 > 0,$$

$$\text{Space-like interval: } (ct)^2 - x^2 = (ct')^2 - x'^2 < 0,$$

$$\text{Null or light-like interval: } (ct)^2 - x^2 = (ct')^2 - x'^2 = 0.$$

5.4.1 *Time machines?*

In General Relativity, there may be "time machines" based on (hypothetical) wormholes in curved spacetime (Mielke, 1977) and foam-like "topological fluctuations" at the tiny Planck scale of $l_p = 10^{-33}$ cm.

These Einstein–Rosen bridges (Fig. 5.16) are unstable and their throats collapse too rapidly before a radio signal can be transmitted through the wormhole. Speculations that exotic (dark?) matter with negative pressure could keep these bridges open are rather unrealistic. Tunnel effects would need some valid theory of quantum gravity.

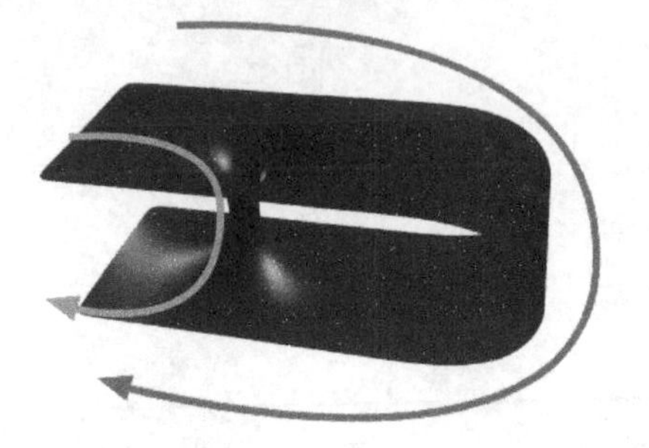

Figure 5.16: Representation of a hypothetical "short-cut" in a wormhole topology.

5.5 Homework

1. Illustrate by a spacetime diagram the phenomenon of contraction of length on a measuring stick, assuming that:

   (a) The observer $\mathcal{O}$ at rest is the one with the measuring stick, and the $\mathcal{O}'$ observer is the one who sees it pass in front of him.

   (b) The observer $\mathcal{O}'$ is moving and is the one with the measuring stick, and the $\mathcal{O}$ observer at rest is the one who sees it pass in front of him.

Solution:

(a) In the first case, if the observer at rest is the one with the rod of length L_0, the world lines of the two ends of the measuring stick are two parallel vertical lines projected upward as shown by the spacetime diagram in Fig. 5.17.

In this case, the observer $\mathcal{O}$ at rest measures for the rod in its proper time $t = 0$ a length L_0. But the observer $\mathcal{O}'$ in motion measures each end of the thin rod at different times and concludes that there was a length contraction of the stick.

(b) In the second case, if the moving observer $\mathcal{O}'$ is the one carrying with him the L_0 measuring rod, the world lines of its two ends remain as two lines which are parallel to their vertical axis ct' as shown in the spacetime diagram of Fig. 5.18. In this case, the observer $\mathcal{O}'$ measures a length L_0 for the stick in its time $ct' = 0$. But the observer $\mathcal{O}$ at rest realizes the coordinates of each end of the rod at different times and concludes that there was a length contraction of the rod.

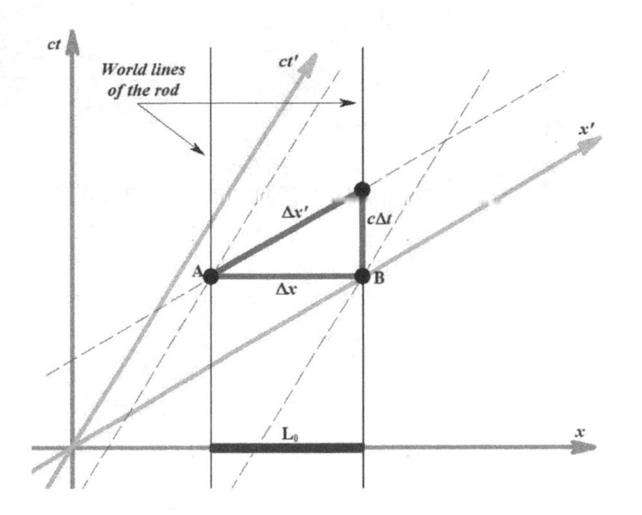

Figure 5.17: The observer $\mathcal{O}$ at rest measures a length L_0 for the rod at the same time $t = 0$, and concludes that there was a length contraction of the stick, with the dots indicating the propagation of the mobile system.

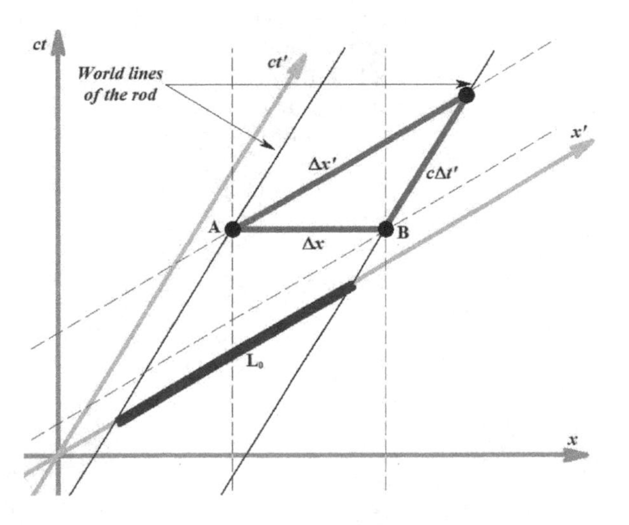

Figure 5.18: The observer $\mathcal{O}'$ measures a length L_0 for the rod at his proper time $t' = 0$. Note that, in the stationary system $\mathcal{O}$, there was a Lorentz contraction.

2. Draw the Minkoswki diagram for the space station. Find coordinates for L'_1 and L'_2.
3. Plot the Lorentz factor γ and the Lorentz contraction L/L_0 as a function of the relative dimensionless velocity $\beta = v/c$.

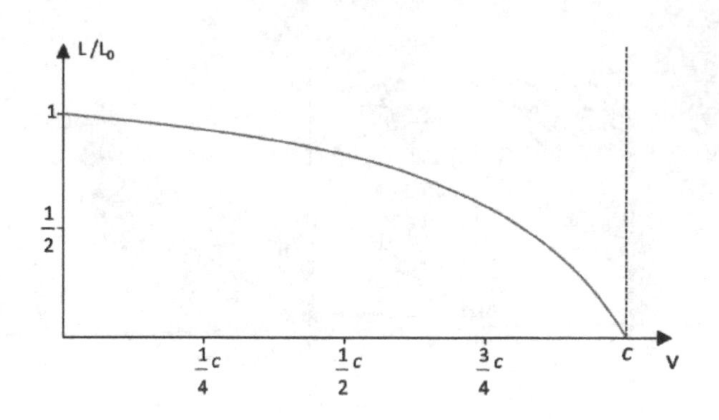

Figure 5.19: Lorentz contraction as a function of the relative velocity.

References

Giulini, D. (2001). "Uniqueness of simultaneity", *British J. Philos. Sci.* **52**, 651–670.

Liebscher, D.E. (1977). *Relativitätstheorie mit Zirkel und Lineal* (Friedr. Vieweg & Sohn, Braunschweig).

Mielke, E.W. (1977). "Knot wormholes in geometrodynamics?", *Gen. Rel. Grav.* **8**, 175–196 [reprinted in: *Knots and Applications*, L.H. Kauffman, ed. (World Scientific, Singapore 1995), pp. 229–250].

Von Laue, M. (1948). "Dr. Ludwig Lange (1863–1936). Ein zu Unrecht Vergessener", *Naturwissenschaften* **35**(7), 193–196.

Chapter 6

Rapidity

Already standard Newtonian mechanics reveals some sort of "relativity".

6.1 Galilean Relativity

Consider a bus trip from Mexico City to a point in the state of Puebla just 100 km from the city. Suppose that the bus maintains a constant linear velocity of $v = 100\,\text{km/h}$. If a person at rest in Mexico City, or alternatively in Puebla, observes the bus, his line of events has the form as shown in Fig. 6.1.

However, viewed from the inertial frame of the bus, Puebla is approaching with a constant opposite velocity till the center of the town "arrives" at the bus. Here the event line tilts backward. In such spacetime diagrams, time runs "up", as in Feynman diagrams.

Then we define the coordinates of the new inertial frame as

$$I' = \begin{cases} \text{axis } t' : x' = 0 = x - vt - x_0 & \text{simultaneous events} \\ \text{axis } x' : t' = 0 = t & \text{(Newton: time is absolute,} \\ & \text{as the emperors of his time).} \end{cases}$$

From these definitions of the axes result **the special transformations of Galilei** (see Fig. 6.2):

$$t' = t$$

$$x' = x - vt - x_0$$

$$u' = u - v.$$

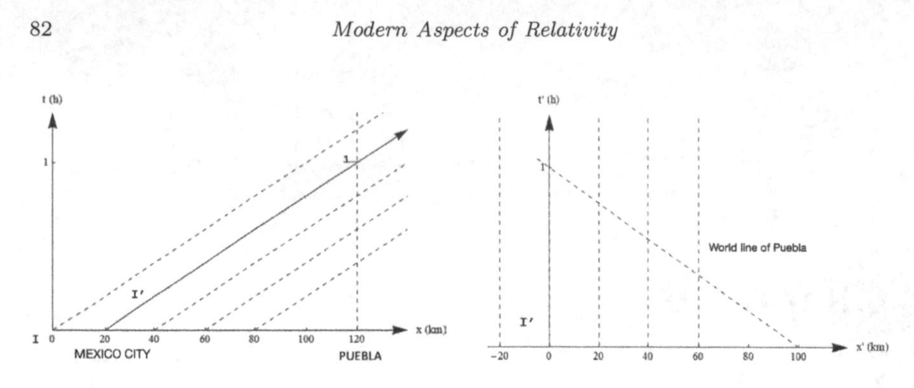

Figure 6.1: Special Galilean transformations.

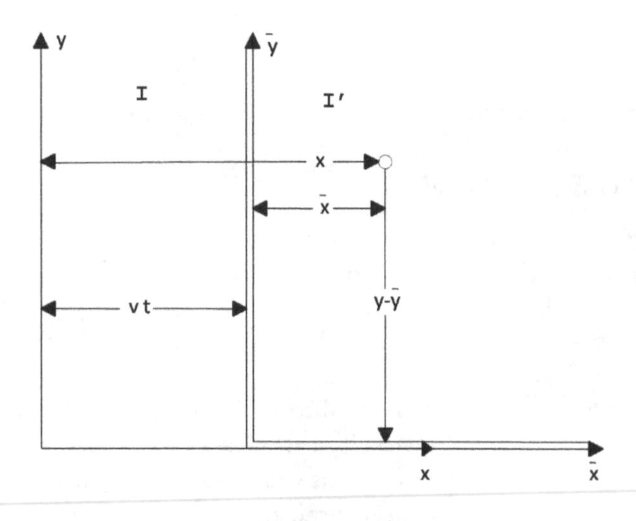

Figure 6.2: Special transformations.

Moreover, one can infer that the instantaneous velocities $u \equiv \dfrac{dx}{dt}$ and $u' \equiv \dfrac{dx'}{dt'}$ are added linearly.

The rules for the Galilean transformations form a rather complicated (inhomogeneous) group with no limits on the velocity, as on German motorways, which is briefly summarized in Appendix A.

6.2 Lorentz Transformations

Suppose an event that occurs at some point P is reported by two observers, one at rest in an inertial frame I and the other at I' moving to the right with a relative velocity v concerning I.

The equations that are valid for all velocity magnitudes $v \leq c$ are

$$\begin{aligned} t' &= \gamma\left(t - \frac{v}{c^2}x\right) \\ x' &= \gamma\,(x - vt) \\ y' &= y \\ z' &= z, \end{aligned}$$

where $\gamma = 1/\sqrt{1 - v^2/c^2}$. These were created[1] by Hendrik A. Lorentz (1853–1928).

When $v \ll c$, these Lorentz transformations reduce to the Galilean transformations:

$$\begin{aligned} t' &\cong t \\ x' &\cong x - vt \\ y' &= y \\ z' &= z \end{aligned}$$

of Newtonian mechanics. Let us note that time was "absolute" for Newton, like the kings and rulers of his epoch.

In many situations, we would like to know the difference between two events P and Q seen by the observers O and O' in different inertial frames as shown in Fig. 6.3.

The Lorentz transformations from one frame of reference to another are

$$I \longrightarrow I' \quad \begin{aligned} \Delta t' &= \gamma\left(\Delta t - \frac{v}{c^2}\,\Delta x\right) \\ \Delta x' &= \gamma\,(\Delta x - v\Delta t), \end{aligned}$$

or

$$I' \longrightarrow I \quad \begin{aligned} \Delta t &= \gamma\left(\Delta t' + \frac{v}{c^2}\,\Delta x'\right) \\ \Delta x &= \gamma\,(\Delta x' + v\Delta t'), \end{aligned}$$

the latter being a Lorentz transformation corresponding to an opposite relative velocity $v \to -v$.

[1] Previously, these transformations were considered in "embryonic" form by Woldemar Voigt in 1887 and probably by Bernhard Riemann who developed in 1858 the wave equation for the electric potential already in its relativistic invariant form. H. Poincaré discussed the mathematics of these transformations in 1904 and the problem of clock synchronization, but could not "liberate" himself from the concept of an absolute frame ("aether").

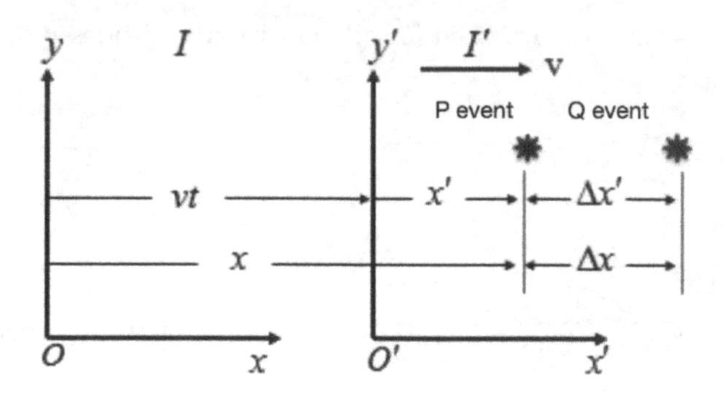

Figure 6.3: Events in two inertial frames I and I'.

From these transformations, it can be deduced that:

(a) For events in the same place, that is $\Delta x = 0$, we have

$$\Delta t' = \gamma \Delta t,$$

which is known as time dilation due to the Lorentz factor.

(b) For simultaneous events, which satisfy $\Delta t = 0$ (in the frame I' in motion relative to I) we recover

$$\Delta x = \frac{\Delta x'}{\gamma} = \sqrt{1 - \frac{v^2}{c^2}} \Delta x' \leq \Delta x'$$

which is the Lorentz (–FitzGerald) contraction, cf. Bell & Weaire (1992). From an object seen from a frame I whose relative velocity v is close to the velocity of light c, its linear length approaches zero. (This does not apply for $v > c$ reasons of causality.)

6.3 Rapidity in 2D

Two individuals or robots in relative motion with respect to each other observe the movement of a body. Suppose the center of the body has an instantaneous velocity component $u'_x = dx'/dt'$ in the direction of the coordinate x.

Since u'_x is measured in the frame I', infinitesimal nearby events obey

$$dx' = \gamma\,(dx - v\,dt)$$
$$dt' = \gamma \left(dt - \frac{v}{c^2} dx \right).$$

The quotient of these (infinitesimal) differentials gives us

$$u'_x = \frac{dx'}{dt'} = \frac{dx - v\,dt}{dt - \frac{v}{c^2}dx} = \frac{\frac{dx}{dt} - v}{1 - \frac{v}{c^2}\frac{dx}{dt}}.$$

Since dx/dt is precisely the component of the velocity u_x of the center of mass in the frame I, this expression becomes

$$u'_x = \frac{u_x - v}{1 - \frac{u_x v}{c^2}}.$$

If such an object has velocity components along the axes y and z, the "transverse" velocities measured by an observer at I' appear also to be transformed:

$$u'_y = \frac{u_y}{\gamma\left(1 - \frac{u_x v}{c^2}\right)},$$

$$u'_z = \frac{u_z}{\gamma\left(1 - \frac{u_x v}{c^2}\right)}.$$

In the non-relativistic case, these components remain invariant and we obtain the Newtonian limit: $u'_x \approx u_x - v$. In the ultra-relativistic case, when $u_x = c$, the equation becomes

$$u'_x = \frac{c - v}{1 - \frac{cv}{c^2}} = \frac{c\left(1 - \frac{v}{c}\right)}{1 - \frac{v}{c}} = c.$$

Thus, Special Relativity (SR) is respecting the invariance of the velocity of light in different frames of a reference, even for a moving laser pointer.

Another consequence of the limit c of the velocity of light for the addition of velocities is that there are no rigid bodies or incompressible fluids in SR; otherwise, instantaneous effects could be transmitted with infinite velocities.

In the studies on relativistic collisions of particles carried out in high-energy laboratories such as CERN, the use of a pseudo-velocity or "rapidity" is common, defined as

$$r \equiv c \tanh^{-1}\left(\frac{v}{c}\right).$$

For engineering, it has the advantage that it can be linearly summed also in SR: That is,

$$R = r_1 + r_2$$

for collinear velocities due to the addition theorem

$$\tanh(\alpha+\beta)=\frac{\tanh\alpha+\tanh\beta}{1+\tanh\alpha\tanh\beta}$$

of the hyperbolic tangent.

Example: Relative velocity of two spaceships

Two spaceships A and B are moving in opposite directions. An observer on Earth measures that the velocity of spacecraft A is 0.750 c and the velocity of spacecraft B is $0.850c$. Find the velocity of ship B as observed by the crew of ship A.

Solution: Identifying the velocities $u_x=-0.850c$ and $v=0.750c$, now we can calculate the velocity u'_x of ship B relative to ship A:

$$u'_x=\frac{u_x-v}{1-\dfrac{u_xv}{c^2}}=\frac{-0.850c-0.750c}{1-\dfrac{(-0.850c)(0.750c)}{c^2}}=-0.977c.$$

Note that the negative sign indicates that ship B is moving in the opposite direction x as observed by the crew of ship A.

6.4 Generalized Lorentz Transformations

For isotropic spaces, the generalized Lorentz transformations (Sexl & Urbantke, 2001) are

$$t'=\gamma\left(t-\frac{\vec{v}\cdot\vec{x}}{v_{\max}^2}\right),$$

$$\vec{x}'=\vec{x}+\frac{\gamma-1}{v^2}\left(\vec{v}\cdot\vec{x}\right)\vec{v}-\gamma\vec{v}t,$$

where $\gamma=1/\sqrt{1-\vec{v}\cdot\vec{v}/v_{\max}^2}$ is the Lorentz factor "generalized" for an arbitrary velocity limit $v_{\max}$. These transformations can be deduced using group theoryy as indicated in Appendix A.

General transformations of Lorentz	Inverse Lorentz transformations
$t'=\gamma\left(t-\dfrac{\vec{v}\cdot\vec{x}}{v_{\max}^2}\right)$	$t=\gamma\left(t'+\dfrac{\vec{v}\cdot\vec{x}'}{v_{\max}^2}\right)$
$\vec{x}'=\vec{x}+\dfrac{\gamma-1}{v^2}\left(\vec{v}\cdot\vec{x}\right)\vec{v}-\gamma\vec{v}t$	$\vec{x}=\vec{x}'+\dfrac{\gamma-1}{v^2}\left(\vec{v}\cdot\vec{x}'\right)\vec{v}+\gamma\vec{v}t'$

The inverse formulae are a result of parity invariance, i.e. $\mathbb{P}:\vec{v}\to\vec{v}'=-\vec{v}$.

Special cases:

(a) Special Galilei transformations:

$$v_{\max} \to \infty \qquad \begin{aligned} t' &= t \\ \vec{x}' &= \vec{x} - \vec{v}\,t. \end{aligned}$$

(b) Let us recover the Lorentz transformations in the direction x:

$$v_{\max} \to c$$

If $\vec{\mathbf{v}} \parallel x$ or $\vec{\mathbf{v}} = (v,\, 0,\, 0)^{\mathrm{T}} \Rightarrow \vec{\mathbf{v}} \cdot \vec{\mathbf{x}} = vx$, then

$$\begin{aligned} t' &= \gamma\left(t - \frac{v}{c^2}x\right) \\ x' &= \gamma\,(x - vt) \\ y' &= y, \quad z' = z, \end{aligned}$$

where $\gamma = 1/\sqrt{1-\beta^2}$ and $\beta = v/c$ is an adimensional velocity.

Transformations	**Inverse transformations**
$t' = \gamma\left(t - \frac{v}{c^2}x\right)$	$t = \gamma\left(t' + \frac{v}{c^2}x'\right)$
$x' = \gamma\,(x - vt)$	$x = \gamma\,(x' + vt')$
$y' = y \qquad z' = z$	$y = y', \qquad z = z'$

6.5 Velocity: The General Lorentz Transformations

Recall that in the Newtonian limit, the vector addition of the velocities $\vec{u}$ and $\vec{v}$ is:

$$\vec{w} = \vec{u} + \vec{v}.$$

Let I, I', and I'' be three inertial frames in relative motion. We want to deduce the generalization of vector addition for relativistic velocities. To this end, we start with the inverse Lorentz transformations as follows:

$$\begin{aligned} t &= \gamma\left(t' + \frac{\vec{v}\cdot\vec{x}'}{c^2}\right) \\ \vec{x} &= \vec{x'} + \frac{\gamma - 1}{v^2}\left(\vec{v}\cdot\vec{x'}\right)\vec{v} + \gamma\vec{v}t'. \end{aligned}$$

Substitute first $\vec{x}' = \vec{u}t'$ and then divide it by t to get $\vec{w} = \vec{x}/t$. Thus, we obtain the rather involved formula:

$$\vec{w} = \frac{\dfrac{\vec{u}}{\gamma} + \dfrac{(\gamma - 1)}{\gamma v^2} (\vec{u} \cdot \vec{v}) \vec{v} + \vec{v}}{1 + \dfrac{\vec{v} \cdot \vec{u}}{c^2}}.$$

An important consequence is the quadratic inequality

$$\vec{w}^2 = c^2 - \frac{c^2 (c^2 - \vec{u}^2) (c^2 - \vec{v}^2)}{(c^2 + \vec{v} \cdot \vec{u})^2} = c^2 \left[1 - \frac{(c^2 - \vec{u}^2) (c^2 - \vec{v}^2)}{(c^2 + \vec{v} \cdot \vec{u})^2} \right] \leq c^2.$$

That is, the squared velocity is always less than or equal to the velocity of light squared, consistent with Einstein's postulates. A more symmetric and equivalent formula (Rindler, 1991) is

$$c^2 - \vec{w}^2 = \frac{c^2 (c^2 - \vec{u}^2) (c^2 - \vec{v}^2)}{(c^2 + \vec{v} \cdot \vec{u})^2}$$

In the ultra-relativistic case, that is, when $|\vec{v}| \simeq c$ and $|\vec{u}| \simeq c$, there results, although counter intuitive, $|\vec{w}| \simeq c$ (see Fig. 6.4).

Special cases:

(a) For parallel velocities $\vec{u} \parallel \vec{v}$, that is $(\vec{u} \cdot \vec{v}) \propto v^2$, we have

$$\vec{w} = \frac{\vec{u} + \vec{v}}{1 + \dfrac{\vec{u} \cdot \vec{v}}{c^2}}.$$

If $|\vec{v}| = c$ y $|\vec{u}| = c$, it implies that $|\vec{w}| = \dfrac{2c}{2} = c$, which, again, is the ultra-relativistic case.

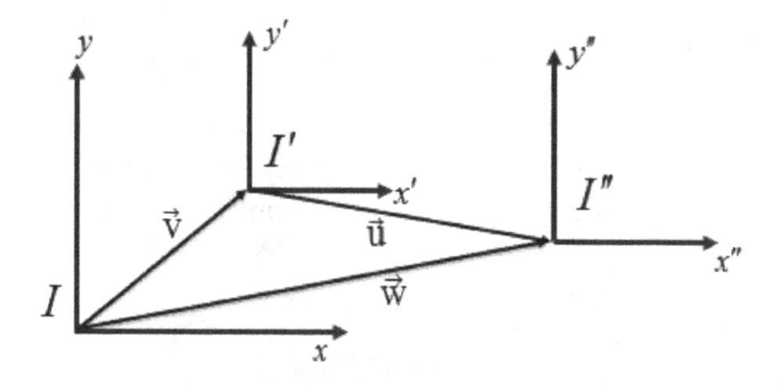

Figure 6.4: Relativistic addition of velocities.

(b) For perpendicular velocities $\vec{u} \perp \vec{v}$, that is $(\vec{u} \cdot \vec{v}) = 0$, the general formula simplifies to

$$\vec{w} = \vec{v} + \sqrt{1 - \frac{v^2}{c^2}}\,\vec{u}.$$

This result can also be obtained as a consequence of time dilation. Moreover, for $|\vec{v}| = c$, there results $|\vec{w}| = |\vec{v}| = c$.

(c) Suppose that the relative velocity $\vec{v}$ has only one component in the direction of x, that is, $\vec{v} = (v, 0, 0)^{\mathrm{T}}$. Since $\vec{u} \cdot \vec{v} = u_x v$, the general form is simplified as in the following table:

Velocity transformations	**Inverse velocity transformations**
$w_x = \dfrac{u_x - v}{1 - u_x v/c^2}$	$w_x = \dfrac{u'_x + v}{1 + v u'_x/c^2}$
$w_y = \dfrac{u_y}{\gamma\,(1 - u_x v/c^2)}$	$w_y = \dfrac{u'_y}{\gamma\,(1 + v u'_x/c^2)}$
$w_z = \dfrac{u_z}{\gamma\,(1 - u_x v/c^2)}$	$w_z = \dfrac{u'_z}{\gamma\,(1 + v u'_x/c^2)}$

6.5.1 *Aberration*

To study the aberration of light as an example of relativistic optics, consider the case that the velocity vector $\vec{u}$ and the axis x' form an angle, more precisely

$$\vec{u} \cdot \vec{v} = u_x v \cos\theta'.$$

If θ' is the angle between $\vec{u}$ and x' as seen from the laboratory frame, we have the following transformation (see Fig. 6.5):

$$\tan\theta = \frac{w_y}{w_x} = \frac{\dfrac{u'_y}{\gamma\,(1 + v u'_x/c^2)}}{\dfrac{u'_x + v}{1 + v u'_x/c^2}} = \frac{u'_y}{\gamma\,(u'_x + v)} = \frac{u' \sin\theta'}{\gamma\,(u' \cos\theta' + v)}.$$

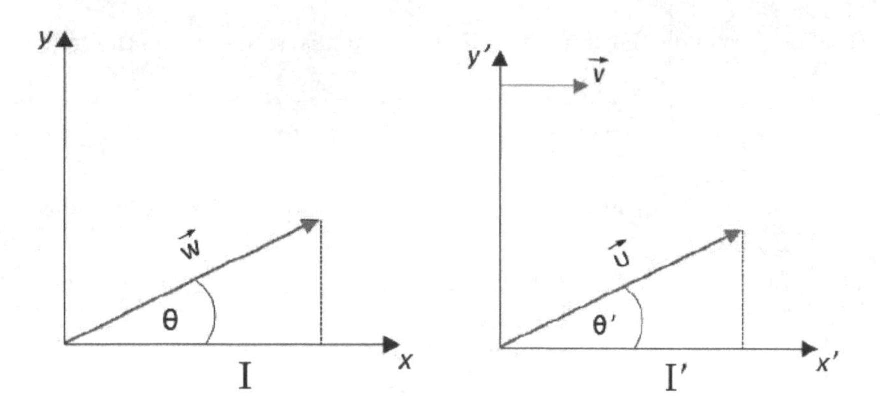

Figure 6.5: Angle θ in different inertial frames.

For the aberration of light, we consider the propagation of photons: If $u' = c$, there results

$$\tan\theta = \frac{\sin\theta'}{\gamma\left(\cos\theta' + \beta\right)}.$$

A relation that will be useful later is the transformation of $\cos\theta$ and $\sin\theta$, which we will determine from the expression of $\tan\theta$. These equations will allow us to deduce the relativistic effects of light emission. Applying trigonometric identities we obtain:

$$\begin{aligned}\tan^2\theta &= \frac{\sin^2\theta'}{\gamma^2\left(\cos\theta' + \beta\right)^2}\\ &= \frac{\sin^2\theta'\left(1 - \beta^2\right)}{\left(\cos\theta' + \beta\right)^2}\\ &= \frac{\sin^2\theta' - \beta^2\sin^2\theta'}{\left(\cos\theta' + \beta\right)^2}\\ &= \frac{1 - \cos^2\theta' - \beta^2 + \beta^2\cos^2\theta'}{\left(\cos\theta' + \beta\right)^2}.\end{aligned}$$

Here the identities $1/\gamma^2 = 1 - \beta^2$ and $\sin^2\theta' = 1 - \cos^2\theta'$ have been applied. Adding one on both sides of the last equations, we have:

$$\tan^2\theta + 1 = \frac{\left(1 + \beta\cos\theta'\right)^2}{\left(\cos\theta' + \beta\right)^2}.$$

From the trigonometric identity $\tan^2\theta' + 1 = 1/\cos^2\theta'$, there results for the inverse:

$$\cos\theta = \frac{\cos\theta' + \beta}{1 + \beta\cos\theta'}.$$

6.5.1.1 *Searchlight effect*

For the transformation of $\sin\theta$, let us apply the trigonometric identity $\sin\theta = \tan\theta\cos\theta$ and proceed as follows:

$$\begin{aligned}\sin\theta &= \tan\theta\cos\theta \\ &= \frac{\sin\theta'}{\gamma(1 + \beta\cos\theta')}.\end{aligned}$$

In the case of photons moving at an angle $\sin\theta' = \pi/2$, $\sin\theta = 1/\gamma$ results. If $\gamma \gg 1$, we can make the approximation $\theta \approx 1/\gamma$.

Even if the photon emission in the inertial frame I′ is isotropic, in the moving frame I it will be "compressed" in the same direction of the movement of the light source. The emission detected in the frame I will be observed as a cone of angular aperture $1/\gamma$ for $\gamma \gg 1$. With increasing velocity v, its degree of collimation will be more notable, see Figs. 6.6 and 6.7.

6.6 Fizeau Interferometer

Originally, Fizeau's experiment was carried out in 1851 to detect the supposed aether and analyze the partial "drag" of light by the flow of water.

Initially, the experiment tried to differentiate between the theory of the static aether and that of the luminiferous aether carried by the medium in which light propagates. If the water moving at velocity v did not "drag along" the aether at all, there should be no difference between the velocity of light in the water at rest c_w or moving respect to the rest frame. If the water "drags along" the aether, the measured velocity should be $c_w + v$ or $c_w - v$ depending on the direction of the water's flow. Surprisingly, the result that Fizeau obtained using interferometry was inconsistent with either model. The amount 0.435 coincides with $(1 - c_w^2/c^2) = 0.4346$ and $c_w = 0.7518\,c$ for water, where c is the velocity of light in vacuum. For other fluids with different velocities of light propagation, the coincidence was similar. The measured velocity was $c_w + v(1 - c_w^2/c^2)$ and $c_w - v(1 - c_w^2/c^2)$ depending on the direction of light propagation relative to $\vec{v}$. Let us see some details of the experiment.

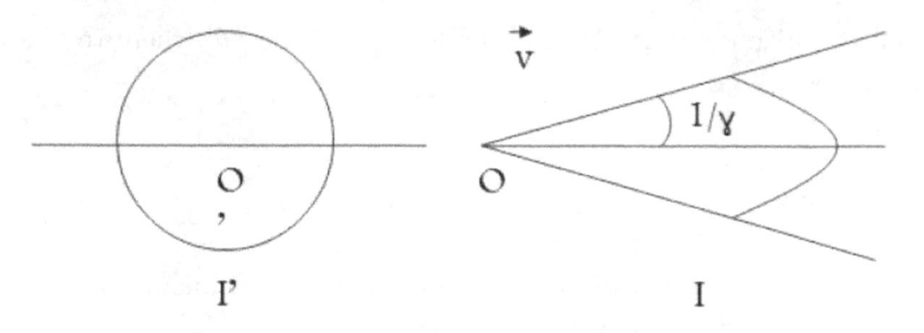

Figure 6.6: "Compressed" radial light emission as a function of the relative velocity.

Figure 6.7: Searchlight effect for waves.

When the water is at rest, the two rays must arrive at the same time because the paths are identical, but when the water is circulating through the transparent tube at velocity v, the red ray will be decelerated and the blue ray will be dragged along.

The displacement of the observed interference fringes coincided, within its precision, with the theory of partially "dragged" photons along the aether. Consequently, the velocities of light along tubes with moving water were

$$c' = \frac{c}{n} \pm v\, d, \quad \text{where } d \equiv \left(1 - \frac{1}{n^2}\right)$$

is the Fresnel (1788–1827) drag coefficient, n being the index of refraction of the fluid. But without the hypothesis of the dragged aether, the experiment was left unexplained. With Einstein and his velocity addition theorem, a complete explanation emerged.

6.6.1 *"Aether drag" experiment in the undergraduate lab*

In Fizeau's experiment, the approximate setup was the following: $L = 1.5\,\text{m}$, $n = 1.33$, $\lambda = 532\,\text{nm}$, and the velocity $v \simeq 7\,\text{m/s}$ of circulating water. An initial shift of 0.23 of the fringes was observed for $v = 0$. Calculate the drag coefficient and compare it with the theoretically obtained value. The time for beam 1 to pass through the water is

$$t_1 = \frac{2L}{(c/n) - v\,d}$$

and for beam 2 is

$$t_2 = \frac{2L}{(c/n) + v\,d}.$$

Consequently,

$$\Delta t = t_1 - t_2 = \frac{4Lv\,d}{(c/n)^2 - v^2\,d^2} \simeq \frac{4Ln^2 v\,d}{c^2}.$$

The period of light of wave length λ is $T = \lambda/c$ so that

$$\Delta N \simeq \frac{\Delta t}{T} = \frac{4Ln^2 v\,d}{\lambda c}$$

and, with the indicated values, we obtain

$$d = \frac{\lambda c \Delta N}{4Ln^2 v} = 0.47.$$

Fresnel's prediction is

$$d = 1 - \frac{1}{n^2} = 0.44.$$

With an inexpensive diode laser, this experiment has recently been realized in an undergraduate laboratory, see Lahaye *et al.* (2012).

6.6.2 *Drag coefficient from Einstein's addition of velocities*

Let us recall the relativistic addition of velocities:

$$w = \frac{u \pm v}{1 \pm \dfrac{u\,v}{c^2}}.$$

Here u and v are the velocities to be added (one is the velocity of the moving frame and the other is the relative velocity between the two reference systems) and w the resulting velocity to be observed.

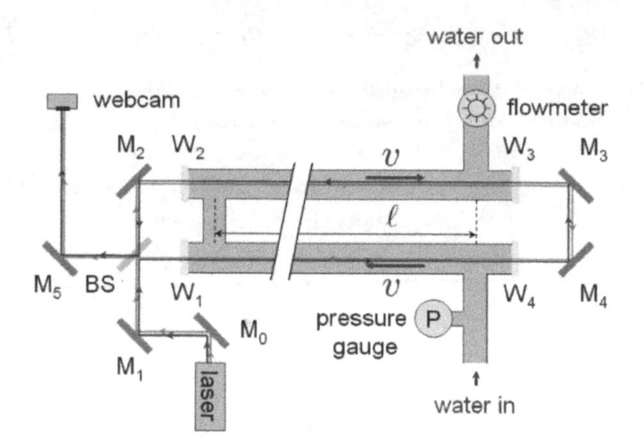

Figure 6.8: Aether theory would tell us that the expected velocities are $c' = c/n \pm v$. The result Fizeau obtained is $c' = c/n \pm (1 - \frac{1}{n^2})v$.

Comparing the formulae with that of a partially dragged aether, it can be seen that the results are almost the same for low velocities of the circulating water:

Following Von Laue (1907), let us substitute u by the velocity c/n of light in water and then substitute this for the section $M_2 - M_3$ of the mirrors in Fig. 6.8:

$$w = \frac{\dfrac{c}{n} \pm v}{1 \pm \dfrac{v\dfrac{c}{n}}{c^2}}.$$

The dependence on quadratic terms of the refraction index is revealed after amplifying the fraction by its algebraic complement:

$$w = \frac{\dfrac{c}{n} \pm v}{\left(1 \pm \dfrac{v\dfrac{c}{n}}{c^2}\right)} \frac{1 \mp \dfrac{v\dfrac{c}{n}}{c^2}}{\left(1 \mp \dfrac{v\dfrac{c}{n}}{c^2}\right)} = \frac{\dfrac{c}{n} \pm v \mp \dfrac{v}{n^2} - \dfrac{v^2}{n\,c}}{\left(1 - \dfrac{v^2}{n^2c^2}\right)}.$$

Then for $v \ll c$, the first-order approximation

$$w \simeq \frac{c}{n} \pm v\left(1 - \frac{1}{n^2}\right)$$

involves Fresnel's coefficient.[2] As a numerical example of the similarity between the expression proposed by Fizeau and that of the relativistic addition of velocities, let us suppose that the velocity of water is $v = 0.001\,c$, and the velocity of light in water $u = 0.75180\,c$. Then for the Fizeau expression a velocity $c' = w = 0.75223480\,c$ is obtained and $c' = w = 0.75223447\,c$ according to the exact velocity addition formula. The difference is not appreciated until reaching the seventh decimal for water velocities one-thousandth of that of light ($c \simeq 300\,000\,\mathrm{km/s}$).

Thus, we can conclude that Einstein's theory explains Fizeau's experiment, which was originally designed to verify the Fresnel "drag" coefficient.

6.7 Homework

1. Suppose a particle is moving at a velocity u' with respect to the frame I$'$. The particle moves in the $x' - y'$ plane and its path forms an angle θ' with the x axis.

   (a) Show that the equations of motion in I$'$ are given by

   $$x' = u't'\cos\theta' \quad y' = u't'\sin\theta' \quad z = 0.$$

   (b) Show that in the frame I, the corresponding velocity u and the angle θ are given by the equations

   $$x = u\,t\,\cos\theta \quad y = u\,t\,\sin\theta \quad z = 0.$$

   (c) Use the Lorentz transformations to verify that the magnitude and direction of the velocity in I are given by

   $$u^2 = \frac{u'^2 + v^2 + 2u'v\cos\theta' - \left(u'^2v^2/c^2\right)\sin^2\theta'}{\left[1 + (u'v/c^2)\cos\theta'\right]^2},$$

   $$\tan\theta = \frac{u'\sin\theta'\sqrt{1-\beta^2}}{u'\cos\theta' + v}.$$

2. Demostrate that

   (a) when $u'^2 = u_x'^2 + u_y'^2$ and $u^2 = u_x^2 + u_y^2$, the quadratic invariant is

   $$c^2 - u^2 = \frac{c^2\left(c^2 - u'^2\right)\left(c^2 - v^2\right)}{\left(c^2 + u_x'v\right)^2}.$$

[2] If light suffers from dispersion $n = n(\omega)$, the refraction factor $1 - 1/n^2$ picks up the additional term $(\omega/n)\, dn/d\omega$.

(b) From this result, show that if $u' < c$ and $v < c$, then u must be less than c. That is, with the relativistic sum of two velocities less than c, one will get a velocity that is also less than c.

(c) From this result, show that if $u' = c$, or $v = c$, then u must equal c. That is, the relativistic sum of any velocity plus the velocity of light is equal to the velocity of light itself.

3. Suppose there is a universe in which the velocity of light is $c = 160\,\text{km/h}$; there a Ferrari travels at the velocity v with respect to a fixed radar equipment: This car surpasses a Volkswagen that travels at the velocity of $80\,\text{km/h} = c/2$. The velocity of the Ferrari is such that the fixed observer measures its length as equal to that of the Volkswagen. If the proper length of the Ferrari is twice that of the Volkswagen, how much does its velocity exceed the velocity limit?
4. Show that proper time, $d\tau = dt\sqrt{1-\beta^2}$, does not vary with respect to the Lorentz transformations. (*Hint*: In the definition of $\beta^2 := v^2/c^2$, substitute $v^2 = (dx/dt)^2 + (dy/dt)^2 + (dz/dt)^2$.)
5. Using the formulae for adding two different velocities $\vec{u}$ and $\vec{v}$, prove the relationship (Rindler, 1991):

$$\gamma(w) = \gamma(u)\gamma(v)\left[1 + \vec{v}\cdot\vec{u}/c^2\right],$$

where $\gamma(u) = 1/\sqrt{1-\vec{u}\cdot\vec{u}/c^2}$ and $\gamma(v) = 1/\sqrt{1-\vec{v}\cdot\vec{v}/c^2}$ are the Lorentz factors for u and v, respectively.
6. Equivalently, for the norm squared of adimensional velocities, show that

$$\vec{\beta}^2 = \frac{(\vec{\beta}_1 + \vec{\beta}_2)^2 - (\vec{\beta}_1 \times \vec{\beta}_2)^2}{(1 + \vec{\beta}_1 \cdot \vec{\beta}_2)^2}$$

holds. *Hint*: Employ identities for the vector product or computer algebra (CA)!

References

Lahaye, T. *et al.* (2012). "Fizeau's "aether-drag" experiment in the undergraduate laboratory", *Am. J. Phys.* **80**, 497.

Von Laue, M. (1907). "Die Mitführung des Lichtes durch bewegte Körper nach dem Relativitätsprinzip", *Ann. Phys.* **328**, 989–990.

Chapter 7

Relativistic Kinematics

7.1 Four-Dimensional Cartesian Coordinates

In Minkowski's spacetime, the three spatial dimensions and time are unified through a four-dimensional (4D) coordinate system composed of

$$x^\mu = \{ct, \vec{x}\} = \{ct, x, y, z\} = (x^0, x^1, x^2, x^3)^{\mathrm{T}},$$

where the curly brackets stand-in for the transposed of a row vector in 4D.

Let us summarize important spacetime concepts:

7.1.1 *Invariant distance*

As a time dilation effect, we obtained earlier $d\tau = dt/\gamma = dt\sqrt{1-\vec{v}^2/c^2}$. Multiplying it by itself gives $d\tau^2 = dt^2 - d\vec{x} \cdot d\vec{x}/c^2$, representing the Pythagorean theorem for a non-Euclidian geometry:

$$ds^2 = c^2 d\tau^2 \stackrel{*}{=} c^2 dt^2 - dx^2 - dy^2 - dz^2 \stackrel{*}{=} \eta_{\mu\nu}\, dx^\mu\, dx^\nu .$$

Here $\eta_{\mu\nu} \equiv \mathrm{diag}(1, -1, -1, -1)$ is a symmetric diagonal matrix, the Minkowski metric.

From now onwards, we will use "Einstein's convention", in which the repetition of two indexes means summation, for example:

$$a_\mu\, b^\mu = \sum_{\mu=0}^{3} a_\mu\, b^\mu .$$

7.1.2 *Proper time*

The time-like infinitesimal distance is defined as

$$d\tau \equiv \frac{ds}{c} = \sqrt{dt^2 - \frac{dx^2 + dy^2 + dz^2}{c^2}}.$$

It does not change by a Lorentz transformation. Then,

$$d\tau = \sqrt{dt^2\left(1 - \vec{\beta}^2\right)} = \frac{dt}{\gamma(v)}, \quad \vec{\beta} \equiv \frac{d\vec{x}}{c\,dt} = \frac{\vec{v}}{c}$$

where $\gamma(v) = \left(1 - \vec{v} \cdot \vec{v}/c^2\right)^{-1/2}$ is the Lorentz factor and $\vec{\beta}$ is an adimensional vector obtained by the division of the velocity of a particle by the celerity of light.

7.1.3 *Time dilation*

On the basis that the Lorentz factor is always $\gamma(v) \geq 1$, we get the following inequality between the differentials of time and proper time:

$$dt = \gamma(v)\, d\tau \geq d\tau.$$

Time dilation is symmetric under an exchange of two inertial frames I and I' with opposite velocities $\vec{v}$ and $\vec{v}' = -\vec{v}$. This is known as parity transformation $\mathbb{P}$ in elementary particle physics.

7.2 Four-D Velocities

Let us define:

$$u^\mu \equiv \frac{dx^\mu}{d\tau} = \frac{dx^\mu}{dt}\frac{dt}{d\tau} = \frac{dx^\mu}{dt}\frac{1}{\sqrt{1-\beta^2}}.$$

The four-vector u^μ represents the relativistic velocity and is a 4D vector of Lorentz. The reason is that if these coordinates are transformed under the Lorentz group, u^μ is transformed in the same way as dx^μ, that is from $dx'^\mu = \Lambda^\mu_\gamma\, dx^\gamma$ results $u'^\mu = \Lambda^\mu_\gamma\, u^\gamma$. Consequently, proper time is an *invariant*, i.e. $d\tau' = d\tau$.

The components of u^μ are

$$u^\mu : \begin{cases} u^0 = \dfrac{c}{\sqrt{1-\beta^2}}, & u^1 = \dfrac{v_x}{\sqrt{1-\beta^2}}, \\[2ex] u^2 = \dfrac{v_y}{\sqrt{1-\beta^2}}, & u^3 = \dfrac{v_z}{\sqrt{1-\beta^2}}. \end{cases}$$

More concisely, one can also write

$$u^{\mu} = \gamma \{c, \vec{v}\}.$$

At rest, $u^{\mu} = \{c, 0, 0, 0\}$ has only a component in the direction of the time axis.

7.2.1 *Quadratic invariant*

Using Minkowski's metric $\eta_{\mu\,\nu}$, we obtain the quadratic identity

$$\eta_{\mu\nu}\, u^{\mu}\, u^{\nu} = \frac{c^2}{1-\beta^2} - \frac{\vec{v}\cdot\vec{v}}{1-\beta^2} = c^2.$$

Since c is constant in any inertial frame, we got a quadratic *invariant* with respect to Lorentz transformations. And the velocity norm is also constant, $|u| = \sqrt{u_{\mu}u^{\mu}} = c$ where $u_{\mu} \equiv \eta_{\mu\nu}\, u^{\nu}$. Therefore, "Special Relativity" is rather a theory invariant with respect to the Lorentz group defined in Appendix A. For the "infrared limit", that is, when $\beta \to 0$, again $u^{\mu} = \{c, 0, 0, 0\}$ is obtained. The Newtonian limit would correspond to $c \to \infty$.

7.3 Acceleration in 4D

Even in SR, an acceleration four-vector can be defined as

$$a^{\mu} \equiv \frac{du^{\mu}}{d\tau} = \frac{du^{\mu}}{dt}\frac{dt}{d\tau} = \frac{du^{\mu}}{dt}\frac{1}{\sqrt{1-\beta^2}}.$$

Considering that

$$\begin{aligned}\dot{\gamma} \equiv \frac{d\gamma}{dt} &= \left(1-\beta^2\right)^{-3/2}\, \vec{\beta}\cdot\frac{d\vec{\beta}}{dt}\\ &= \gamma^3\, \vec{\beta}\cdot\frac{d\vec{\beta}}{dt},\end{aligned}$$

the time derivative of the absolute value of the relative velocity is

$$\frac{d|\,\vec{v}\,|}{dt} = \frac{d\left(\sqrt{v_x^2+v_y^2+v_z^2}\right)}{dt} = \frac{d\left(\sqrt{\vec{v}\cdot\vec{v}}\right)}{dt} = \frac{1}{|\,\vec{v}\,|}\left(\vec{v}\cdot\frac{d\vec{v}}{dt}\right).$$

Then, for the Cartesian components of acceleration, we get the following explicit expressions:

$$a^0 = \frac{\vec{v}\cdot\vec{a}}{c\,(1-\beta^2)^2}, \qquad a^1 = \frac{dv_x/dt}{1-\beta^2} + \frac{v_x\,(\vec{v}\cdot\vec{a})}{c^2\,(1-\beta^2)^2},$$

$$a^2 = \frac{dv_y/dt}{1-\beta^2} + \frac{v_y\,(\vec{v}\cdot\vec{a})}{c^2\,(1-\beta^2)^2}, \quad a^3 = \frac{dv_z/dt}{1-\beta^2} + \frac{v_z\,(\vec{v}\cdot\vec{a})}{c^2\,(1-\beta^2)^2}.$$

Since $\vec{a} = d\vec{v}/dt$ is the Newtonian acceleration, a more concise but equivalent form is

$$a^\mu = \gamma \left\{ c\frac{d\gamma}{dt}, \frac{d\gamma}{dt}\vec{v} + \gamma\vec{a} \right\}$$
$$= \gamma^2\{\gamma^2\vec{\beta}\cdot\vec{a},\ \vec{a} + \gamma^2(\vec{\beta}\cdot\vec{a})\vec{\beta}\}.$$

Consequently, when the acceleration is perpendicular to the velocity, i.e. for $\vec{\beta}\cdot\vec{a} = 0$, it reduces to

$$a^\mu = \gamma^2\{0, \vec{a}\}.$$

In a reference frame, where $\vec{\beta} = 0$, $a^\mu = \{0, \vec{a}\}$ depends only on the Newtonian acceleration.

In general, as

$$\eta_{\mu\nu}\, a^\mu\, a^\nu = -\left[c^2\,\dot{\gamma}^2 + \gamma^4\,\vec{a}\cdot\vec{a}\right] \le 0,$$

the 4D acceleration is always a space-like vector. Departing from the quadratic equation $\eta_{\mu\nu}\, u^\mu\, u^\nu = c^2$, its derivative with respect to proper time yields:

$$\frac{d}{d\tau}\left(\eta_{\mu\nu}\, u^\mu\, u^\nu\right) = 0 = 2\,\eta_{\mu\nu}\,\frac{du^\mu}{d\tau}\,u^\nu = 2\,\eta_{\mu\nu}\,a^\mu\,u^\nu$$

Thus, in Special Relativity acceleration a^μ is always "perpendicular" to the velocity u^ν in 4D.

Due to the principle of equivalence, acceleration and gravitation are *locally* indistinct in General Relativity (Einstein's elevator or microgravitation in a spacecraft). This principle has been experimentally verified by measuring, via "lunar laser ranging", very precisely the distance to the moon (Williams *et al.*, 2012).

7.4 Relativistic Momentum

Similar to the Newtonian linear momentum, for which $\vec{p} \cong m\vec{v}$ holds, in relativity, the 4D momentum for a free particle is defined as

$$p^\mu \equiv m\, u^\mu.$$

It is called the relativistic momentum, and

$$p^0 = mu^0 = \frac{mc}{\sqrt{1-\beta^2}} = \frac{1}{c}E$$

is directly proportional to the energy E, multiplied by the factor $1/c$. This factor is necessary according to the physical unit of energy in the international system that is the Joule ($1\,\mathrm{J} \equiv 1\,\mathrm{kg\,m^2/s^2}$).

7.5 Quadratic Invariant for the Momentum

Similarly to the quadratic indentity for the velocity in 4D, the quadratic momentum is defined by

$$\begin{aligned}\frac{E^2}{c^2} - \vec{p}\cdot\vec{p} &= \eta_{\mu\nu}\, p^\mu\, p^\nu \\ &= m^2 \eta_{\mu\nu}\, u^\mu\, u^\nu \\ &= m^2 c^2\end{aligned}$$

which is independent of the inertial frame. It is invariant with respect to the transformations of Lorentz since the velocity of light remains constant. Then mass is also an invariant characteristic of any particle. A "relativistic mass" concept leads to a contradiction (Okun, 1989). Some attempts to quantize gravitation would lead to additional terms in this dispersion relation. However, they violate Lorentz's invariance and have not been corroborated (Popelov & Romalis, 2004). Clearing the energy in the above formula, we get Einstein's relationship:

$$\boxed{E = \pm c\,\sqrt{\vec{p}^{\,2} + m^2c^2}}\,,$$

where we can see that $E \neq mc^2$.

By convention, the positive sign is reserved for a particle and the opposite sign corresponds to its anti-particle. In the Stueckelberg–Feynman interpretation, energy is always positive. However, there is a reflection $\mathbb{T}$ of time and a change $\mathbb{C}$ of charge, which subsequently lead to pairs such as the electron and the positron. In 2010, an entire nucleus of anti-matter of Helium-4 consisting of two antiprotons and two antineutrons was formed by collisions (STAR Collaboration, 2011).

- **Special cases for E:**

(a) Photon,[1] $m = 0$, $E = cp = c|\vec{p}|$.
(b) Energy at rest, $|\vec{p}| = 0 \Rightarrow$

[1] In a medium with refractive index n there is a controversy, between Minkowski and Abraham, regarding the correct formulation for photons. In vacuum, that is when $n = 1$, both formulae match, see Barnett (2010).

In quantum mechanics, E and $\vec{p}$ are replaced by operators $E \to i\hbar\partial/\partial t$ and $\vec{p} \to -i\hbar\vec{\nabla}$ acting on the wave function φ . Then the relativistic invariant $p^2 \equiv \eta_{\mu\nu}\, p^\mu\, p^\nu$ converts into the Klein–Gordon equation

$$\Box\varphi + \frac{m^2c^2}{\hbar^2}\varphi = 0,$$

where $\Box = \dfrac{\partial^2}{c^2\partial t^2} - \vec{\nabla}\cdot\vec{\nabla}$ is the 4D wave operator and $\hbar = h/2\pi$ the reduced Planck constant. For photons the mass is $m = 0$, otherwise there would be a discontinuity in the gravitational deviation of light rays near the sun, which has not been observed (see Goldhaber & Nieto, 2010).

7.6 Relativistic Energy Corrections

If we substitute the relativistic formula for the impetus $\vec{p} = m\vec{v}/\sqrt{1-\beta^2}$ in the energy E, we recover the equivalent formula:

$$E = mc^2\sqrt{\frac{\beta^2}{1-\beta^2} + 1} = \frac{mc^2}{\sqrt{1-\beta^2}} = \gamma mc^2.$$

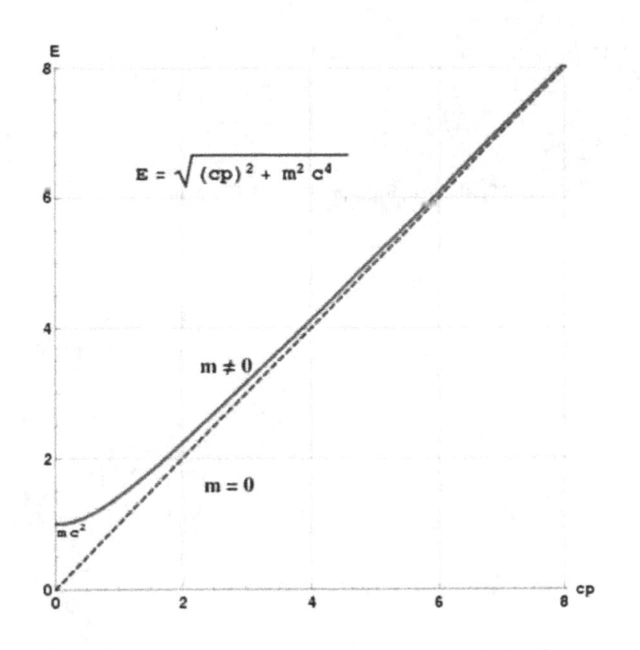

Figure 7.1: Relativistic energy and the "mass shell" (red) for $m \neq 0$.

For small velocities and therefore for $v \ll c$ one can expand the previous formula in a Taylor series, which is an expansion of a smooth function into an infinite sum of terms:

$$E = mc^2 + \frac{1}{2}mv^2 + \frac{3}{8}mv^2\left(\frac{v}{c}\right)^2 + \frac{5}{16}mv^2\left(\frac{v}{c}\right)^4 + \cdots$$

Physically, it corresponds to:

$E \approxeq$ **energy at rest + kinetic energy + relativistic corrections.**

The kinetic energy can be expanded as well:

$$E_{\text{kinetic}} \equiv E - E_0 = \frac{1}{2}mv^2\left[1 + \frac{3}{4}\beta^2 + \frac{5}{8}\beta^4 + \cdots\right].$$

where the common factor is familiar from Newtonian mechanics. (As all drivers should be aware of, the kinetic energy grows at least *quadratically* with velocity.)

7.7 Relativistic Shape of Mechanics

In Newtonian mechanics, force can be obtained by the temporal derivative of the linear impetus, that is by Newton's second law. In the same manner, but using the proper time, we can define Minkowski's force in 4D as

$$K^{\mu} \equiv \frac{d}{d\tau} p^{\mu} = \frac{d}{d\tau} (m u^{\mu}) .$$

Since the mass of a point particle is invariant, we find

$$K^{\mu} = m a^{\mu}.$$

Here K^{μ} is the four-vector of Minkowski's force, $u^{\mu} = \{c/\sqrt{1-\beta^2}, \vec{v}/\sqrt{1-\beta^2}\}$ the velocity in 4D and $a^{\mu} = du^{\mu}/d\tau$ the relativistic acceleration.

Therefore, the spatial components with $i = 1, 2, 3$ are

$$m\frac{du^i}{d\tau} = m\frac{du^i}{dt}\frac{dt}{d\tau} = \frac{m}{\sqrt{1-\beta^2}} \frac{d\left(v^i/\sqrt{1-\beta^2}\right)}{dt} = K^i.$$

That can also be represented in vector form as

$$\frac{d}{dt}\left(\frac{m\vec{v}}{\sqrt{1-\beta^2}}\right) = \vec{K}\sqrt{1-\beta^2} \equiv \vec{F}.$$

In relativity, the vector force is obtained by deriving the relativistic relation $\vec{p} = m\gamma\vec{v}$ and using the fact that the mass of a particle is an invariant. Observe that, in SR, the 3D force depends explicitly on the velocity

$$\begin{aligned} \vec{F} \equiv \frac{d\vec{p}}{dt} &= m\left(\gamma\vec{a} + \dot{\gamma}\vec{v}\right) \\ &= m\left[\gamma\vec{a} + \gamma^3\left(\vec{a}\cdot\vec{\beta}\right)\vec{\beta}\right] \\ &= m\gamma\left[\vec{a} + \gamma^2\,\frac{\vec{v}\cdot\vec{a}}{c^2}\,\vec{v}\right]. \end{aligned}$$

- **Special cases for $\vec{F}$:**

(a) $\vec{F} \perp \vec{v} \;\Rightarrow\; \vec{a} = \dfrac{\vec{F}}{m\gamma}$ where $m_{\perp} = m\gamma$ is, at times, called "transverse mass".

(b) $\vec{F} \parallel \vec{v} \;\Rightarrow\; \vec{a} = \dfrac{\vec{F}}{m\gamma}\left(1 - \vec{\beta}\cdot\vec{\beta}\right) = \dfrac{\vec{F}}{m\gamma^3}$ where $m_{\parallel} = m\gamma^3$ is, at times, called "longitudinal mass". However, the only correct concept is that of the invariant mass m.

These special cases can be summarized in matrix form as

$$\begin{pmatrix} F_{\parallel} \\ F_{\perp} \end{pmatrix} = m \begin{pmatrix} \gamma^3 & 0 \\ 0 & \gamma \end{pmatrix} \begin{pmatrix} a_{\parallel} \\ a_{\perp} \end{pmatrix},$$

These are the only cases where the force is proportional to the acceleration.

Let us now check the (1+3)D decomposition

$$ma^{\mu} = K^{\mu} \iff \begin{pmatrix} \dfrac{dE}{dt} = \vec{F} \cdot \vec{v} \\ \dfrac{d\vec{p}}{dt} = \vec{F} \end{pmatrix}.$$

When deriving the invariant $\eta_{\mu\nu} p^{\mu} p^{\nu} = m^2 c^2$ with respect to proper time, we see that the Minkowski force is also perpendicular to the impetus

$$\eta_{\mu\nu} K^{\mu} p^{\nu} = 0 = K^0 p^0 - \vec{K} \cdot \vec{p}.$$

Clearing for the time component K^0 and eliminating the non-vanishing invariant mass, we get

$$K^0 = \frac{1}{p_0} \vec{K} \cdot \vec{p} = \frac{c}{E} \vec{K} \cdot \vec{p} = \frac{1}{c} \frac{\vec{F} \cdot \vec{v}}{\sqrt{1 - \beta^2}} = \frac{\gamma}{c} \vec{F} \cdot \vec{v}.$$

Thus, Minkowski's force has the following components:

$$K^{\mu} = \left\{ \frac{1}{c} \frac{\vec{F} \cdot \vec{v}}{\sqrt{1 - \beta^2}}, \frac{\vec{F}}{\sqrt{1 - \beta^2}} \right\} = \gamma \{ \vec{F} \cdot \vec{\beta}, \vec{F} \}.$$

The temporal component is, by definition

$$K^0 = m \frac{d\left(c/\sqrt{1 - \beta^2} \right)}{d\tau} = \frac{1}{c} \frac{\vec{F} \cdot \vec{v}}{\sqrt{1 - \beta^2}}$$

or

$$\frac{d}{dt} \left(\frac{mc^2}{\sqrt{1 - \beta^2}} \right) = \vec{F} \cdot \vec{v},$$

where $E = \dfrac{mc^2}{\sqrt{1 - \beta^2}}$ is the equivalent form of the relativistic energy. Thus, we can finally identify:

$$\vec{K} = \gamma \vec{F},$$

$$\vec{p} = \gamma m \vec{v},$$

The appearance of the Lorentz factor in the linear momentum has already been anticipated in previous chapters.

For the scalar product $\vec{F} \cdot \vec{\beta}$, it can be shown that

$$\vec{a} = \frac{\vec{F} - (\vec{F} \cdot \vec{\beta})\vec{\beta}}{m\gamma}.$$

It is important to note that acceleration is not always parallel to the force, in contradiction to Newtonian mechanics. Only in the limit when $\beta \to 0$, which implies $\gamma \to 1$, Newton's formula for the force is retained.

7.8 Spinning Top

In Newtonian mechanics, the vector product

$$L = \vec{x} \times \vec{p}$$

is the angular momentum and

$$\vec{M} = \frac{d\vec{L}}{dt} = \vec{x} \times \vec{F}$$

the moment or torque of a spinning top.

In Special Relativity, angular momentum is a 4D tensor

$$L^{\mu\nu} \equiv x^\mu p^\nu - x^\nu p^\mu.$$

It is anti-symmetric, and its derivative with respect to proper time

$$\begin{aligned} M^{\mu\nu} &\equiv \frac{\partial L^{\mu\nu}}{\partial \tau} \\ &= x^\mu K^\nu - x^\nu K^\mu \end{aligned}$$

preserves the tensor character. The 4D torque is given in terms of Minkowski's force K^μ, since the anti-symmetric products of the four-momenta drop out.

When an extended body rotates, some density cells get accelerated. Due to Einstein's equivalence principle, acceleration is locally equivalent to gravity, macroscopically well described by General Relativity. Similarly as in Mach's principle, local inertial frames are influenced and dragged by the distribution and flow of mass or energy-momentum in the Universe. In particular, they are dragged by the motion and rotation of nearby bodies.

The most extreme "spinning tops" in the Universe are millisecond pulsars, i.e. rotating neutron stars (NS, see Manko *et al.*, 2000) of about 20 km diameter. Spacetime gets curved and other objects in their vicinity experience a "frame dragging", like some torque on a gyroscope.

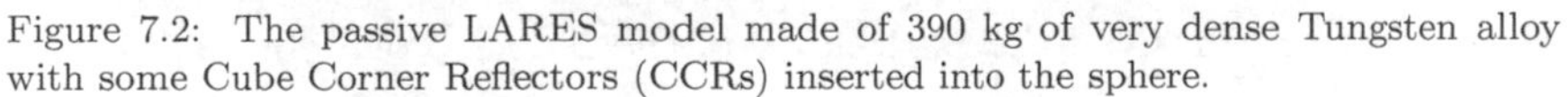

Figure 7.2: The passive LARES model made of 390 kg of very dense Tungsten alloy with some Cube Corner Reflectors (CCRs) inserted into the sphere.

In order to corroborate this (Einstein-) Lense–Thirring effect, passive satellites like the LAser RElativity Satellite (LARES) are lifted into an orbit around the Earth, cf. Fig. 7.2. With laser interferometry, one measures this tiny frame dragging due to the Earth's and the satellite's rotation. Quadrupole and higher moments like in NSs may affect the precision (Ciufolini *et al.*, 2012).

7.9 Group Velocity of a Particle

Albert Einstein anticipated the duality between wave and particles of light (massless photons) already in 1905. In 1923, Louis de Broglie proposed the quantum hypothesis that any particle must also show such a duality. Accordingly, the velocity of a particle should always be equal to the group velocity of the corresponding wave packet. De Broglie deduced that if the concepts of duality already known for light hold-up for any particle, then his hypothesis would be more easily accepted. The Compton effect and the interference experiments with an electron microscope (Tonomura, 2005) confirmed such matter waves.

Defining the group velocity via

$$v_g = \frac{\partial \omega}{\partial k} = \frac{\partial (E/\hbar)}{\partial (p/\hbar)} = \frac{\partial E}{\partial p},$$

where E is the total energy of the particle, p its momentum, and $\hbar = h/2\pi$ the reduced Planck constant. The 4D wave vector $(\omega/c, \vec{k})$ enters here.

In Special Relativity, we find

$$\begin{aligned} v_g = \frac{\partial E}{\partial p} &= \frac{\partial}{\partial p}\left(\sqrt{p^2c^2 + m^2c^4}\right) \\ &= \frac{1}{2}\frac{2pc^2}{\sqrt{p^2c^2 + m^2c^4}} \\ &= \frac{p}{m\sqrt{(p/mc)^2 + 1}}. \end{aligned}$$

If we replace $p = mv\gamma$, this simplifies to

$$v_g = \frac{mv\gamma}{m\gamma} = v.$$

Here m is the mass of the particle, c is the velocity of light in vacuum, $\gamma(v) = 1/\sqrt{1 - v^2/c^2}$ the Lorentz factor, and v is the particle velocity regardless of its behavior as matter waves.

In the non-relativistic limit, we get, for a free particle with kinetic energy $p^2/2m$, in the same manner

$$\begin{aligned} v_g = \frac{\partial E}{\partial p} &\simeq \frac{\partial}{\partial p}\left(\frac{1}{2}\frac{p^2}{m}\right) \\ &= \frac{p}{m} = v. \end{aligned}$$

For a photon with $E = pc$ in vacuum, the "ultra-relativistic" limit $v_g = c$ is retained, as expected. (Anomalous dispersion is not considered here.)

7.10 Homework

1. Show that $\vec{F} \cdot \vec{\beta} = m\,\gamma^3 \vec{a} \cdot \vec{\beta}$ using the identity $\gamma^2 - 1 = \gamma^2 \vec{\beta} \cdot \vec{\beta}$. Compute the difference $\vec{F} - \left(\vec{F} \cdot \vec{\beta}\right)\vec{\beta}$ and use the interim result to demonstrate that

$$\vec{a} = \frac{\vec{F} - (\vec{F} \cdot \vec{v})\vec{v}/c^2}{\gamma m}.$$

2. For collinear velocities and accelerations, i.e. when $\vec{v} \cdot \vec{a} = |\vec{v}||\vec{a}|$, prove that

$$F = |\vec{F}| = m\gamma^3 a,$$

holds, where $a = \sqrt{\vec{a} \cdot \vec{a}}$ is the absolute value of a vector.

3. Derive from the 4D torque $M^{\mu\nu}$ the relativistic equation

$$\frac{d}{dt}(\, m\vec{r} - t\vec{p}\,) = \frac{\vec{r}}{c}(\vec{F} \cdot \vec{\beta}) - t\vec{F}$$

for the center of mass (Ruder, 1993).

References

Ciufolini, I. *et al.* (2012). "Overview of the LARES Mission: Orbit, error analysis and technological aspects", *Phys.: Conf. Ser.* **354**, 012002; *Eur. Phys. J.* **76**, 120.

Goldhaber, A.S. and Nieto, M.M. (2010). "Photon and graviton mass limits", *Rev. Mod. Phys.* **82**, 939–979.

Manko, V.S. *et al.* (2000). "Exact solution for the exterior field of a rotating neutron star", *Phys. Rev.* **61** , 081501 (Rapid Communication).

Okun, L.B. (1989). "The concept of mass", *Phys. Today* **31** (June 1989), 31–36.

Pospelov, M. and Romalis, M. (2004). "Lorentz invariance on trial", *Phys. Today,* July, 40–46.

STAR Collaboration (2011). "Observation of the antimatter helium-4 nucleus", *Nature* **473**, 353–356.

Williams, J.G. *et al.* (2012). "Lunar laser ranging tests of the equivalence principle", *Class. Quant. Grav.* **29**, 184004.

Chapter 8

Relativistic Billiards

8.1 Collision of Identical Particles

Let us consider first the *elastic* collision of two identical particles.[1] In a two-dimensional (2D) plane (like a billiard table) both are of the same mass m. Before the collision, the particle 2 is at rest with respect to the laboratory frame of reference. Therefore, the particles will have the following values for the momentum and energy:

$$1: E \text{ and } \vec{p} \quad \text{and} \quad 2: E_0 = mc^2 \text{ and } \vec{p}_0 = 0.$$

After the collision, their values would be

$$1: E_1 \text{ and } \vec{p}_1 \quad \text{and} \quad 2: E_2 \text{ and } \vec{p}_2.$$

Furthermore, if no external force is applied, due to Noether's theorem:

$$\frac{d}{d\tau} p^\mu = 0.$$

This is known as the law of conservation of 4D relativistic momentum. In our laboratory frame of reference, this means that

$$p_1{}^\mu + p_2{}^\mu = \bar{p_1}{}^\mu + \bar{p_2}{}^\mu \qquad \begin{cases} E + E_0 = E_1 + E_2 \\ \vec{p} + \vec{0} = \vec{p}_1 + \vec{p}_2 \end{cases}$$

Now we apply the invariance of the momentum square ("mass-shell") for the following two particles:

$$\eta_{\mu\nu} p^\mu p^\nu = p_\mu p^\mu = m^2c^2 \quad \textit{and} \quad \bar{p}_\mu \bar{p}^\mu = m^2c^2.$$

[1] Without considering an internal structure like the spin or helicity of a Fermion.

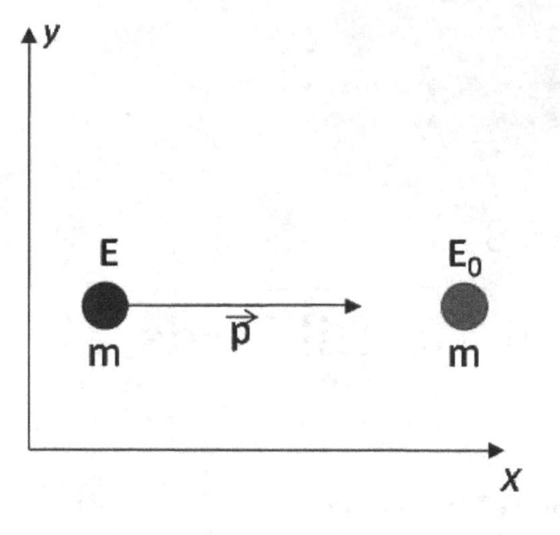

Figure 8.1: Before the impact.

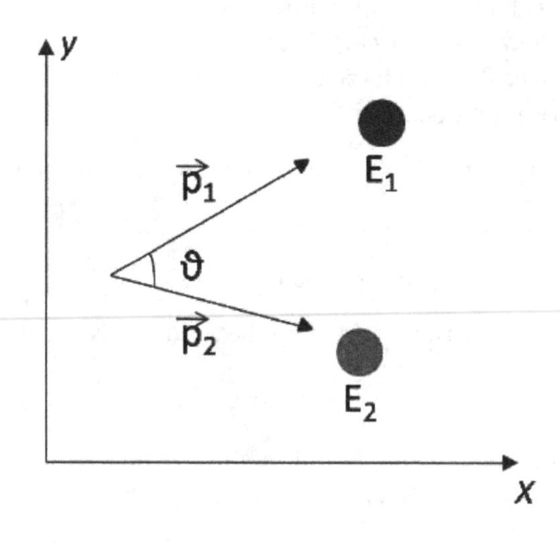

Figure 8.2: After the collision.

As a consequence, the cross terms of the final momenta should satisfy

$$p_{1\mu} p_2^\mu = \bar{p}_{1\mu} \bar{p}_2^\mu$$

which implies that

$$E^2 = c^2(\vec{p} \cdot \vec{p} + m^2 c^2) \Rightarrow |\vec{p}| = \frac{1}{c}\sqrt{E^2 - E_0^2}.$$

In the same manner, we can write

$$|\vec{p}_1| = \frac{1}{c}\sqrt{E_1^2 - E_0^2} \quad \text{and} \quad |\vec{p}_2| = \frac{1}{c}\sqrt{E_2^2 - E_0^2}.$$

Geometrically, we can use for the absolute value $|\vec{p}| = p$ the law of cosine as follows:

$$p^2 = p_1^2 + p_2^2 - 2p_1p_2\cos(\pi - \vartheta) = p_1^2 + p_2^2 + 2p_1p_2\cos\vartheta.$$

Substituting the values of p^2, p_1^2, and p_2^2, we have

$$\frac{1}{c^2}\left(E^2 - E_0^2\right) = \frac{1}{c^2}\left(E_1^2 - E_0^2\right) + \frac{1}{c^2}\left(E_2^2 - E_0^2\right) + 2p_1p_2\cos\vartheta,$$

and, after the addition of the term $2m^2c^2 = 2E_0^2/c^2$ on both sides, we find

$$\frac{E^2}{c^2} + \frac{E_0^2}{c^2} = \frac{E_1^2}{c^2} + \frac{E_2^2}{c^2} + 2p_1p_2\cos\vartheta.$$

Solving for the scattering angle, we get

$$\cos\vartheta = \frac{E^2 + E_0^2 - E_1^2 - E_2^2}{2p_1p_2c^2}.$$

Let us eliminate E_2 and the momenta p_1 and p_2, using the following algebraic relations:

$$\begin{aligned} E_2 &= E + E_0 - E_1 \\ E^2 + E_0^2 - E_1^2 - E_2^2 &= 2\left(E - E_1\right)\left(E_1 - E_0\right) \\ p_1 &= \frac{1}{c}\sqrt{E_1^2 - E_0^2} = \frac{1}{c}\sqrt{\left(E_1 - E_0\right)\left(E_1 + E_0\right)} \\ p_2 &= \frac{1}{c}\sqrt{E_2^2 - E_0^2} = \frac{1}{c}\sqrt{\left(E - E_1\right)\left(E - E_1 + 2E_0\right)}, \end{aligned}$$

such that

$$\cos\vartheta = \frac{1}{\sqrt{\left(1 + \dfrac{2E_0}{E_1 - E_0}\right)\left(1 + \dfrac{2E_0}{E - E_1}\right)}} < 1$$

results.

Note that the momentum vectors are located in an ellipse with major and minor half axes, respectively,

$$a = \frac{p}{2} \quad \text{and} \quad b = \frac{p}{\sqrt{2\left(1 + E/E_0\right)}} \leq \frac{p}{2}$$

for $E \geq E_0$ and $p = |\vec{p}|$. When $E \geq E_1 \geq E_0$ is satisfied, there results the range

$$\vartheta_{\min} < \vartheta \leq \pi/2$$

for the scattering angle. In the Newtonian limit, when $c \to \infty$, we obtain $\cos\vartheta \to 0$, which implies that $\vartheta \to \pi/2$, familiar to an ordinary billiard player.

Limiting case: In the non-relativistic limit, i.e. $v \ll c$, we have $E \approx E_0$ and the ellipse enclosing the momenta will degenerate to the Thales circle[2] with $\vartheta = \pi/2$, as all billiard players know.

For $p_1 = p_2$, Fig. 8.3 tells us that

$$\tan\frac{\vartheta_{\min}}{2} = \frac{b}{a} = \sqrt{\frac{2}{(1 + E/E_0)}}.$$

8.1.1 *Elastic collisions*

Relativistic billiards are "played" today in (rather expensive) particle accelerators such as CERN in Geneva, where central collisions between electrons

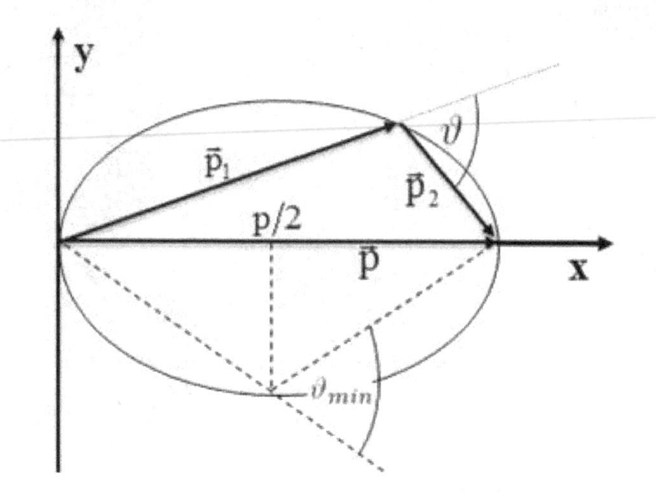

Figure 8.3: Diagram of the kinematics $\vec{p} = \vec{p_1} + \vec{p_2}$ (Ruder, 1993).

[2]Thales of Miletus (624–546 B.C.) is regarded as the father of deductive geometry. A theorem of Thales states: "Every angle inscribed in a semicircle is a right angle."

and/or protons occur. This relativistic problem of collisions is solved more easily when the total momentum is constant:

$$\bar{p} = \sum p_i = \text{constant}.$$

The constancy of the 3D part of the four-vector moment is the generalization of conservation of $\vec{p}$ in Newtonian mechanics, while the constancy of the fourth or temporal component is the relativistic generalization of the conservation of energy. Now, suppose that the two particles can be specified initially by the following four-momenta:

$$p = M\gamma(u)\{c, \vec{u}\} \quad \text{and} \quad q = m\gamma(w)\{c, \vec{w}\}.$$

After an elastic collision we know, by definition, that the number of particles and their respective masses do not change. Therefore, when the sum of $\bar{p}$ is constant, we have

$$p + q = p' + q'$$

or equivalently,

$$M\gamma(u)\{c, \vec{u}\} + m\gamma(w)\{c, \vec{w}\} = M\gamma(u')\{c, \vec{u}'\} + m\gamma(w')\{c, \vec{w}'\},$$

where the primed letters represent amounts after the collision.

Squaring both sides and considering that the relation $p^2 = p'^2 = M^2c^2$ is fulfilled also for q, we find:

$$p \cdot q = p' \cdot q' = p_t q_t - p_x q_x - p_y q_y - p_z q_z.$$

We can use any inertial frame to calculate this invariant. If we consider M as a bullet and m as the blanket, it is natural to choose the frame of reference of m (laboratory frame of reference) where $\vec{w} = 0$. Thus, again

$$p \cdot q = p' \cdot q' = Mm\gamma(u)c^2.$$

However, since this value should be the same after the collision, we realize that the absolute value of u must be invariant. The magnitude $u = |\vec{u}|$ is known as the *relative velocity* and the absolute value of the velocity of one of the particles with respect to the frame of reference of the other. This is the most important result of the elastic collision of two particles: the magnitude of the relative velocity does not change, that is, the vector of the relative velocity can change direction but not its magnitude.

Now we will study the case of a central collision where both velocities are along the line joining the centers of the particles, for example, the x axis.

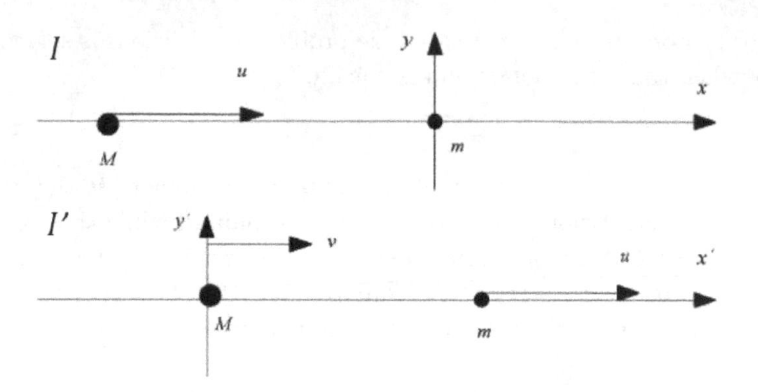

Figure 8.4: Diagrams of collisions; before the collision, the system is seen in I and afterwards is seen in I'. The velocity v in I' is the velocity of M after the collision.

First, we calculate the four-vector of the total momentum in the reference frame I of m and obtain

$$\bar{p} = p + q = \{(M\gamma(u) + m)c,\ M\gamma(u)u,\ 0,\ 0\}.$$

In I' where M is at rest, the total four-vector is

$$\bar{p}\,' = \{(M + m\gamma(u))c,\ m\gamma(u)u,\ 0,\ 0\}.$$

The velocity v of I' with respect to I should be the velocity of M after the collision. The momenta $\bar{p}$ and $\bar{p}\,'$ are not equal, since they refer to a different frame. However, they are related by a Lorentz transformation, which gives us

$$m\gamma(u)u = \gamma(v)\left[M\gamma(u)u - \frac{v}{c}(M\gamma(u) + m)c\right]$$

for the component p'_x and

$$(M + m\gamma(u))c = \gamma(v)\left[(M\gamma(u) + m)c - \frac{v}{c}M\gamma(u)u\right]$$

for the component p'_t. To find the velocity v of M after the collision, we combine the above equations and eliminate the Lorentz factor $\gamma(v)$ to get the following linear relation:

$$m\gamma u\left((M\gamma + m)c - \frac{v}{c}M\gamma u\right) = (M + m\gamma)c\big[M\gamma u - v(M\gamma + m)\big],$$

where $\gamma \equiv \gamma(u)$ is a function of u only. Depending on whether $M > m$ or not, the bullet will continue on its way or otherwise gets deflected and will return. Solving for v, we have:

$$v = v(u) = \frac{(M^2 - m^2)u}{M^2 + m^2 + 2Mm/\gamma(u)}.$$

The non relativistic case is obtained when $\gamma(u) = 1$ and yields

$$v_{\rm N}(u) = \frac{M - m}{M + m}u,$$

known from Newtonian mechanics. In the ultra-relativistic limit, when $u \to c$, we would obtain

$$v(c) = \frac{M^2 - m^2}{M^2 + m^2}c.$$

The ratio of these equations gives us (Essén, 2002):

$$\begin{aligned}\frac{v_{\rm N}}{v(c)} &= \frac{(M - m)(M^2 + m^2)}{(M + m)(M^2 - m^2)}\frac{u}{c} \\ &= \frac{M^2 + m^2}{(M + m)^2}\frac{u}{c} \\ &\simeq \frac{1}{2}\frac{u}{c}.\end{aligned}$$

For $M \simeq m$, as approximately in the case of protons and neutrons, the ultra-relativistic velocity is twice that of Newtonian mechanics for particles with the same mass.

8.2 Mandelstam Variables

These Lorentz invariant variables were introduced 1957 by Stanley Mandelstam after the death of Einstein in 1955. Now, let us consider two particles of masses m_1 and m_2 of any type and the inelastic collision between them which satisfies $p_1 + p_2 = p_3 + p_4$ by four-momentum conservation:

Each of Lorentz invariant combinations of the incoming or outgoing four-momenta p_1^μ, p_2^μ, p_3^μ and p_4^μ can be expressed in terms of the Mandelstam variables (Berkovits *et al.*, 2017):

$$\begin{aligned}s &\equiv (p_1 + p_2)^2 = (p_3 + p_4)^2 \\ t &\equiv (p_1 - p_3)^2 = (p_4 - p_2)^2 \\ u &\equiv (p_1 - p_4)^2 = (p_3 - p_2)^2.\end{aligned}$$

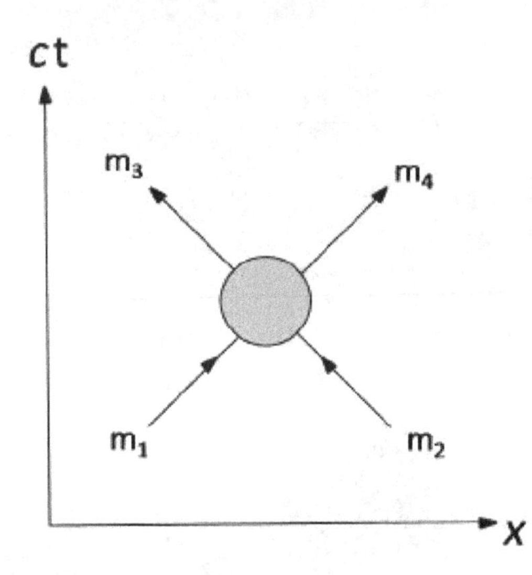

Figure 8.5: Collision of two particles of masses m_1 and m_2.

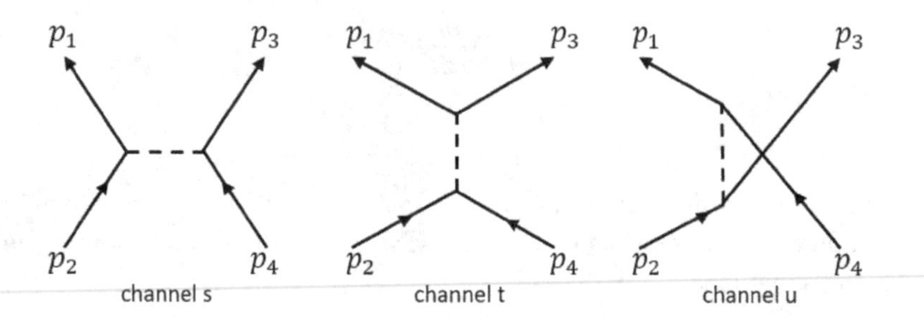

Figure 8.6: Feynman type diagrams of the channels s, t, and u of the Mandelstam variables.

Conceptionally, they provide a more direct representation of the possible channels of particle scattering, since $p^2 \equiv \eta_{\mu\nu} p^\mu p^\nu$ is an invariant with respect to Lorentz transformations, cf. Kibble (1960).

The total energy in the collision "enters" in the variable s and the four-momentum transfer has its counterpart in the variable t, whereas u corresponds to a "crossing symmetry" for the in- and out-going states.

In terms of the Mandelstam variables, the Lorentz products $p_i \cdot p_j \equiv \eta_{\mu\nu} p_i^\mu p_j^\nu$ of four-momenta are given by

$$2(p_1 \cdot p_2) = s - m_1^2 c^2 - m_2^2 c^2,$$

$$2(p_3 \cdot p_4) = s - m_3^2 c^2 - m_4^2 c^2,$$

$$2(p_1 \cdot p_3) = m_1^2c^2 + m_3^2c^2 - t,$$

$$2(p_2 \cdot p_4) = m_2^2c^2 + m_4^2c^2 - t,$$

$$2(p_1 \cdot p_4) = m_1^2c^2 + m_4^2c^2 - u,$$

$$2(p_2 \cdot p_3) = m_2^2c^2 + m_3^2c^2 - u.$$

Consequently, only two of these three variables are independent, since

$$\begin{aligned} s + t + u &= \left[m_1^2 + m_2^2 + m_3^2 + m_4^2\right] c^2 + 2p_1 \cdot [p_1 + p_2 - p_3 - p_4] \\ &= \left[m_1^2 + m_2^2 + m_3^2 + m_4^2\right] c^2. \end{aligned}$$

The second parenthesis is zero due to the conservation of four-momentum and the "on shell" conditions for all p^2 terms. In relativistic thermodynamics, this set of formulae uses averages over the light cone, see Dunkel *et al.* (2009). More recently, Mandelstam variables have been used also in the relativistic Boltzmann equation (see Cerignani & Kremer, 2002, and Strain, 2010).

Let us consider the interaction between two particles of momenta p_1 and p_2 and of masses m_1 and m_2 transforming into particles of momenta p_3 and p_4 and of final masses m_3 and m_4. When colliding, the Lorentz invariant Mandelstam variables can be rewritten, for $c = 1$, as

$$s = m_1^2 + 2E_1E_2 - 2\vec{p_1} \cdot \vec{p_2} + m_2^2$$

$$t = m_1^2 - 2E_1E_3 + 2\vec{p_1} \cdot \vec{p_3} + m_3^2$$

$$u = m_1^2 - 2E_1E_4 + 2\vec{p_1} \cdot \vec{p_4} + m_4^2,$$

and satisfy, again,

$$s + t + u = \left[m_1^2 + m_2^2 + m_3^2 + m_4^2\right] c^2.$$

The cross-section of two body collisions has the rate

$$\frac{d\sigma}{dt} = \frac{1}{64\pi s} \frac{1}{|\vec{p}_{\mathrm{CM}}|^2} |\mathcal{M}|^2,$$

where, in QFT, $\mathcal{M}$ is a matrix invariant under Lorentz transformations.

Example: Particle–antiparticle annihilation

In particular, for electrons and positrons the process of annihilation produces two photons:

$$e^- e^+ \to \gamma\gamma$$

The balance of momenta $p_- + p_+ \to k_1 + k_2$, where k denotes the wave-vector of a photon satisfying $k_\mu k^\mu = 0$, corresponds to

$$\begin{aligned} s + t + u &= 2m_e^2 c^2, \\ 2(p_- p_+) &= s - 2m_e^2 c^2, \\ 2(k_1 k_2) &= s, \\ 2(p_- k_1) = 2(p_+ k_2) &= m_e^2 c^2 - t, \\ 2(p_- k_2) = 2(p_+ k_1) &= m_e^2 c^2 - u. \end{aligned}$$

The differential cross-section is given by

$$d\sigma = \frac{(2\pi)^4 |\mathcal{M}|^2}{4\sqrt{(p_1 \cdot p_2)^2 - m_1^2 m_2^2}} \quad d\Phi_n\,(p_1 + p_2;\; p_3, \ldots, p_{n+2}),$$

in natural units, where $c = 1$. The denominator can be written in the inertial frame of the mass m_2, i.e. in the laboratory frame, as

$$\sqrt{(p_1 \cdot p_2)^2 - m_1^2 m_2^2} = m_2 p_{1\,\mathrm{lab}}$$

while in the reference system of the center of mass (CM), it can be re-written as

$$\sqrt{(p_1 \cdot p_2)^2 - m_1^2 m_2^2} = p_{1\mathrm{CM}} \sqrt{s}.$$

8.2.1 *Center of mass*

Let us consider the collision of two particles, the interaction can be elastic or inelastic when new particles may occur.

Examples

(a) Collision of two protons $\quad p + p \longrightarrow p + p$

(b) Generation of new particles $\quad \begin{cases} p + p \longrightarrow p + p + p + \bar{p} \\ \pi^- + p \longrightarrow k^0 + \Lambda \end{cases}$

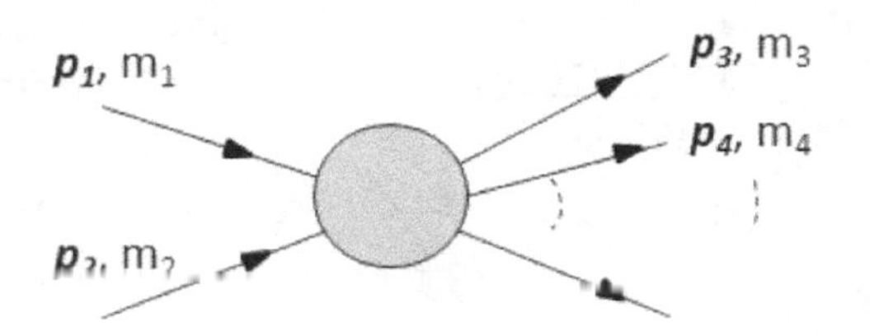

Figure 8.7: Particle production.

Primarily, these frames are distinguished by

(a) Center of mass: $\vec{p_1} + \vec{p_2} = 0$

(b) Laboratory: $\vec{p_2} = 0$

For these two systems, one of the Mandelstam variables is, again for $c = 1$, given by

$$s = (E_1 + m_2)^2 - \vec{p_1}^{\,2} = (E_1^* + E_2^*)^2.$$

Since $E_1^2 = \vec{p_1}^{\,2} + m_1^2$, the total energy in the center of mass frame reduces to

$$\sqrt{s} = E_1^* + E_2^* = \sqrt{m_1^2 + m_2^2 + 2E_1 m_2},$$

The ultrarelativistic case, where $E_1 \gg m_1$ and $E_1 \gg m_2$, results

$$\sqrt{s} \simeq \sqrt{2E_1 m_2},$$

That is, the energy in the center of mass grows as the square root of the energy in the laboratory frame. At CERN, in the search for the Higgs boson, in 2012 a total energy of

$$\sqrt{s} \simeq 8\,\mathrm{TeV}$$

was achieved. For subtleties in the statistical interpretation, see Lyons (2012).

8.3 Inelastic Collision of Particles in the CM Frame

Consider the collision of two particles of different initial masses m_1 and m_2 and final masses m_3 and m_4, after colliding. Let us recall the Mandelstam variables:

$$\begin{aligned} s &= (p_1 + p_2)^2, \\ t &= (p_1 - p_3)^2, \\ u &= (p_1 - p_4)^2. \end{aligned}$$

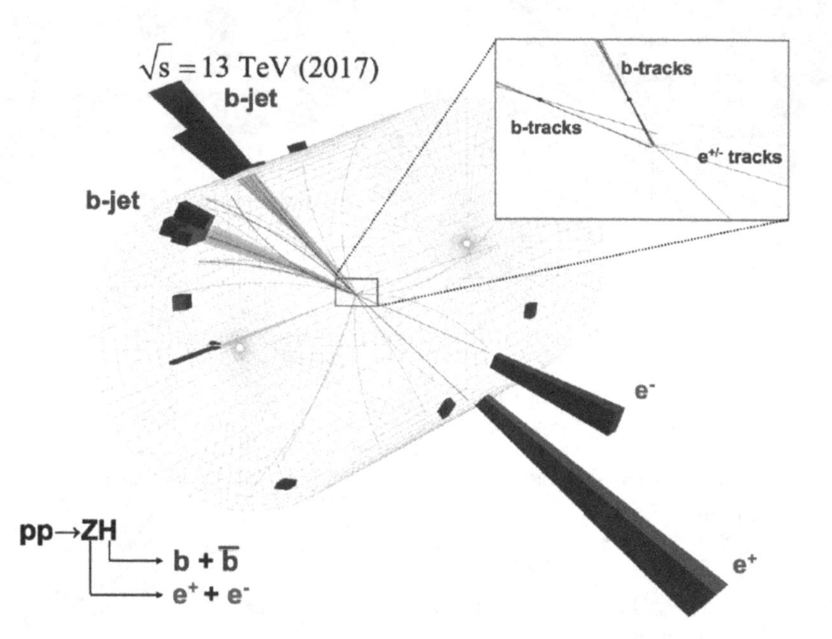

Figure 8.8: A CMS candidate event for the Higgs boson (H) decaying to two bottom quarks (b), in association with a Z boson decaying to an electron (e^-) and a positron (e^+).

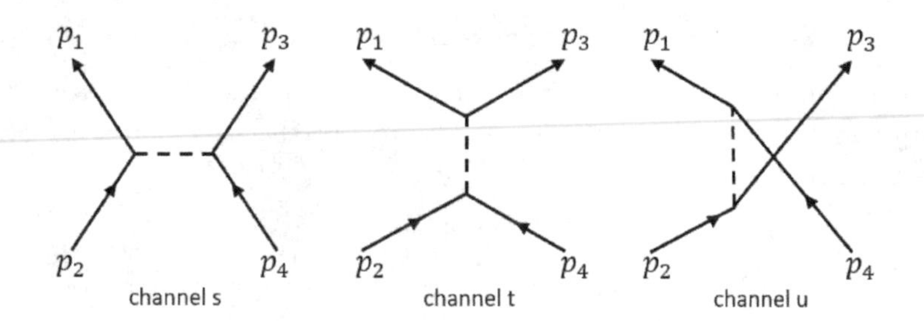

Figure 8.9: Visualization of the Mandelstam variables. The space, time, and u channels represent different Feynman type diagrams.

Again, s (space-channel) denotes the square of the total energy of the center of mass (positive) and t (time-channel) is the square of the transferred four-momentum (negative). Also $s + t + u = \sum_{i=1}^{4} m_i^2$ in natural units where $c = 1$.

The center of mass frame is defined by

$$\vec{p_1} + \vec{p_2} = 0 = \vec{p_3} + \vec{p_4}.$$

The corresponding variables are denoted with an asterisk: (CM. $p_i = p_i^*$). The laboratory frame of reference is defined by $\vec{p_2} = 0$ (fixed) and the variables are tagged with $p_i = p_i^{\text{lab}}$.

The center of mass (CM) frame leads to the following equations:

$$\begin{aligned} \vec{p_1^*} &= -\vec{p_2^*} = \vec{p} \\ \vec{p_3^*} &= -\vec{p_4^*} = \vec{p}' \end{aligned}$$

Consequently, the four-momenta are

$$\begin{aligned} p_1 &= \left(E_1^* = \sqrt{\vec{p}^2 + m_1^2}, \vec{p}\right) \\ p_2 &= \left(E_2^* = \sqrt{\vec{p}^2 + m_2^2}, -\vec{p}\right) \\ p_3 &= (E_3^*, \vec{p}') \\ p_4 &= (E_4^*, -\vec{p}') \end{aligned}$$

Hence, the Mandelstam variable s takes the value:

$$s = (p_1 + p_2)^2 = (E_1^* + E_2^*)^2.$$

Now we can express E_i^*, $|\vec{p}|$, and $|\vec{p}'|$ in terms of s as

$$\begin{aligned} E_{1,3}^* &= \frac{1}{2\sqrt{s}}(s + m_{1,3}^2 - m_{2,4}^2), \\ \vec{p}^{*2} &= (E_1^*)^2 - m_1^2 = \frac{1}{4s}\lambda(s, m_1^2, m_2^2), \end{aligned}$$

where we have employed the Källén triangle function defined by

$$\begin{aligned} \lambda(a,b,c) &= a^2 + b^2 + c^2 - 2(ab + ac + bc) \\ &= \left[a - (\sqrt{b} + \sqrt{c})^2\right]\left[a - (\sqrt{b} - \sqrt{c})^2\right] \\ &= a^2 - 2a(b+c) + (b-c)^2. \end{aligned}$$

Observe that the Källén function has the following properties:

- Symmetric under interchange of the sides of the triangle, $a \leftrightarrow b \leftrightarrow c$ and
- Asymptotic behavior: For $a \gg b, c$: we find $\lambda(a, b, c,) \to a^2$

The "physical region" of s, t, u in the three possible channels is depicted in Fig. 8.10. This allows us to determine some properties of dispersion

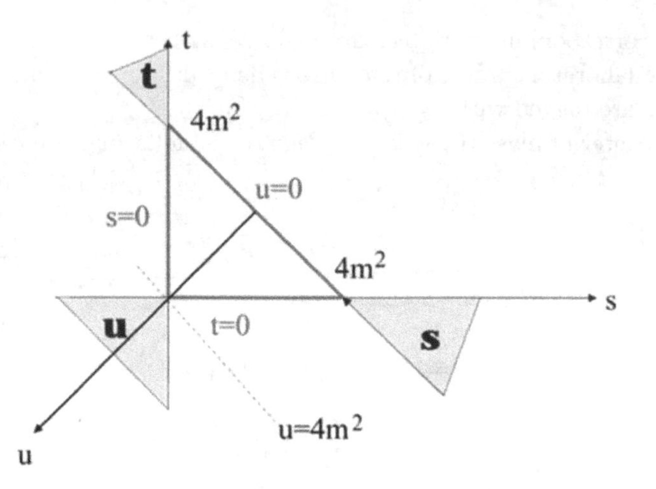

Figure 8.10: Physically restricted regions of the s, t, u channels in the Cartesian st-plane for equal masses m. Every point in the plane satisfies $s + t + u = 4m^2$. (Note that the unit along the u axis is smaller by factor $1/\sqrt{2}$ compared to the t and s axes.)

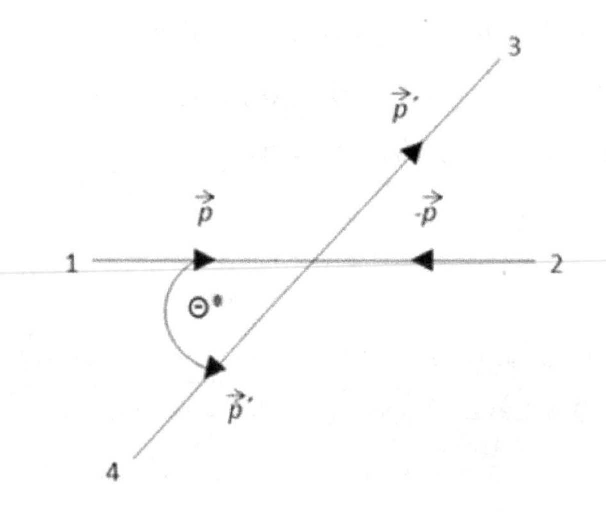

Figure 8.11: Collision of two particles in the center of mass frame. A restriction is the scattering angle Θ^*.

processes. From $\vec{p}^2, \vec{p}^{*2} > 0$, it follows that

$$s_{\min} = \max\left\{(m_1 + m_2)^2, (m_3 + m_4)^2\right\} \geq 0$$

at the beginning of the collision in the channel s. In the limit of high-energies, i.e., $s \gg m_i^2$, the equations of energy and momentum are simplified by the asymptotic behavior of λ and we obtain

$$E_1^* = E_2^* = E_3^* = E_4^* = |\vec{p}| = |\vec{p}^*| = \frac{\sqrt{s}}{2}.$$

Recently, the Higgs boson characterized by $\sqrt{s} = 13$ TeV, while decaying into bottom quarks, produced a two jet event (CERN, 2018).

8.3.1 *Scattering angle*

In the center of mass frame, the scattering angle Θ^* is defined by

$$\vec{p} \cdot \vec{p}' = |\vec{p}| \cdot |\vec{p}'| \cos\Theta^*,$$

Moreover, we also know that

$$\begin{aligned} p_1^\mu p_{3\mu} = p_1 \cdot p_3 = E_1^* E_3^* - |\vec{p_1}^*||\vec{p_3}^*| \cos\Theta^* \\ t = (p_1 - p_3)^2 = m_1^2 + m_3^2 - 2p_1 p_3 \\ = m_1^2 + m_2^2 - 2E_1^* E_3^* + 2\vec{p_1} \cdot \vec{p_3}. \end{aligned}$$

Consequently, we can deduce that $\cos\Theta^*$ is an explicit function of (s, t, m_i^2) involving the Källén triangle function: Solving for $\cos\Theta^*$, we get

$$\cos\Theta^* = \frac{\vec{p_1} \cdot \vec{p_3}}{|\vec{p_1^*}||\vec{p_3^*}|} = \frac{t - m_1^2 - m_3^2 + 2E_1^* E_3^*}{2|\vec{p_1^*}||\vec{p_3^*}|}.$$

When substituting $|\vec{p_1^*}|$ and $|\vec{p_3^*}|$,

$$\cos\Theta^* = \frac{2s(t - m_1^2 - m_3^2 + 2E_1^* E_3^*)}{\sqrt{\lambda(s, m_1^2, m_2^2)}\sqrt{\lambda(s, m_3^2, m_4^2)}}$$

results. Simplifying the numerator, we get

$$\begin{aligned} &2s(t - m_1^2 - m_3^2 + 2E_1^* E_3^*) \\ &= 2s\left[t - m_1^2 - m_3^2 + 2\left(\frac{s + m_1^2 - m_2^2}{4s}\right)\left(\frac{s + m_3^2 - m_4^2}{1}\right)\right] \\ &= 2s(t - m_1^2 - m_3^2) + s^2 + sm_3^2 - sm_4^2 + sm_1^2 + m_3^2 m_1^2 \\ &\quad - m_4^2 m_1^2 - sm_2^2 - m_2^2 m_3^2 + m_2^2 m_4^2 \\ &= 2st - s\,(s + t + u) + s^2 + m_3^2 m_1^2 - m_4^2 m_1^2 - m_2^2 m_3^2 + m_2^2 m_4^2 \\ &= 2st - st - su + m_1^2\,(m_3^2 - m_4^2) - m_2^2\,(m_3^2 - m_4^2). \end{aligned}$$

By factorization, we finally obtain

$$\cos\Theta^* = \frac{s(t-u) + (m_1^2 - m_2^2)(m_3^2 - m_4^2)}{\sqrt{\lambda(s, m_1^2, m_2^2)}\sqrt{\lambda(s, m_3^2, m_4^2)}}.$$

This means that the scattering of two particles $2 \to 2$ is described by two independent variables:

$$\sqrt{s} \text{ and } \Theta^* \quad \text{or} \quad \sqrt{s} \text{ and } t.$$

8.4 Compton Effect

Already in 1923, Arthur H. Compton studied the scattering of X-rays by electrons in a carbon target probing relativity. To measure the angular dependence θ of X-rays as they exit the target, a rotating crystal is used. According to a classical interpretation, we can consider this effect as a result of a collision between a photon and an electron. Therefore, it is sufficient to consider the energy and the relativistic momentum:

Before the collision, according to the "relativistic triangle" previously introduced, we have

$$\vec{E}^2 = (\vec{p} \cdot \vec{p})c^2 + \left(mc^2\right)^2,$$

where $E_1 = |\vec{p_1}|c$ for a photon and $E_0 = m_e c^2$ for an electron approximately at rest in the target. After the collision, $E_2 = |\vec{p_2}|c$ holds for the photon and $E_e^2 = \vec{p_e} \cdot \vec{p_e}\, c^2 + (m_e c^2)^2$ for the scattered electron.

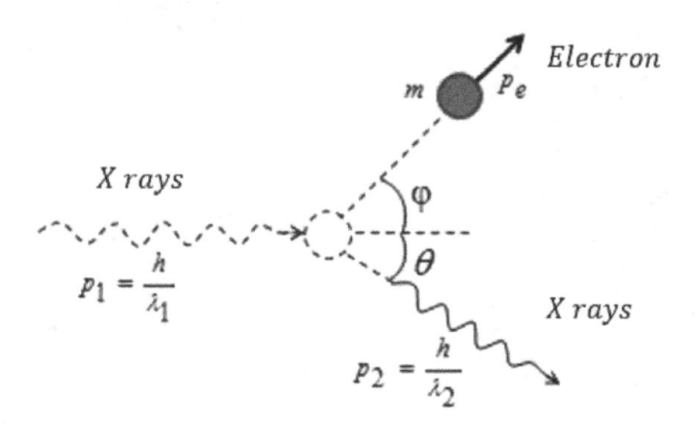

Figure 8.12: Compton effect: Collision between a particle and an X-ray.

On the other hand, the conservation of three-momentum can be expressed as

$$\vec{p_e} = \vec{p_1} - \vec{p_2}.$$

Multiplying each side of this vector equation by itself, we find for the scalar product that

$$\vec{p_e}^{\,2} = \vec{p_1}^{\,2} + \vec{p_2}^{\,2} - 2\vec{p_1}\cdot\vec{p_2} = p_1^2 + p_2^2 - 2p_1p_2\cos\theta.$$

Energy conservation $E_0 + E_1 = E_2 + E_e$ gives us the additional relation:

$$p_1c + m_ec^2 = p_2c + \sqrt{p_e^2c^2 + (m_ec^2)^2}.$$

Solving this equation for the electron momentum p_e, we get

$$\begin{aligned} p_e^2c^2 &= \left[(\vec{p_1} - \vec{p_2})\,c + m_ec^2\right]^2 - \left(m_ec^2\right)^2 \\ &= \left(\vec{p_1} - \vec{p_2}\right)^2 c^2 + 2|\vec{p_1} - \vec{p_2}|m_ec^3, \end{aligned}$$

or, equivalently,

$$\begin{aligned} p_e^2 &= \left(\vec{p_1} - \vec{p_2}\right)^2 + 2(p_1 - p_2)m_ec \\ &= p_1^2 + p_2^2 - 2p_1p_2 + 2m_ec(p_1 - p_2). \end{aligned}$$

Comparing this equation with the conservation of the momentum squared, it follows that

$$p_1^2 + p_2^2 - 2p_1p_2 + 2m_ec\,(p_1 - p_2) = p_1^2 + p_2^2 - 2p_1p_2\cos\theta,$$

After cancellation of identical terms, we obtain

$$m_ec\,(p_1 - p_2) = p_1p_2\,(1 - \cos\theta)\,.$$

Let us multiply both sides by $h/(m_ecp_1\,p_2)$ and, after a partial cancellation, conclude that

$$\frac{h}{p_2} - \frac{h}{p_1} = \frac{h}{m_ec}\,(1 - \cos\theta)\,.$$

Although we introduced the Planck constant h on both sides of the equation, this is still a classic result. However, using de Broglie's wavelength $h/p = \lambda$ for the quantum waves associated with matter, Compton's scattering equation

$$\boxed{\lambda_2 - \lambda_1 = \frac{h}{m_ec}\,(1 - \cos\theta)}$$

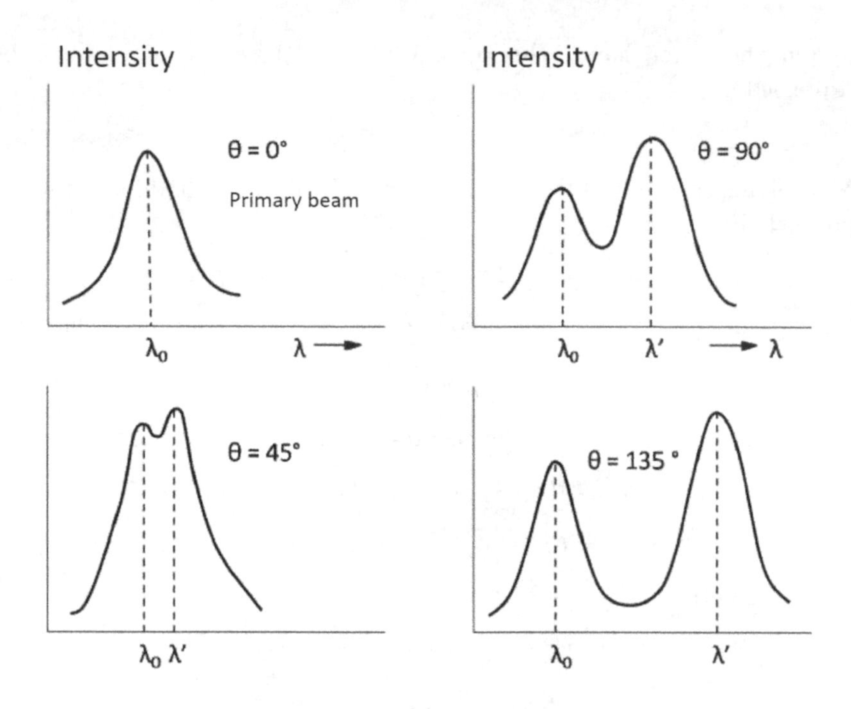

Figure 8.13: X-ray intensity depending on the refractive angle θ.

results. The magnitude

$$\lambda_C = \frac{h}{mc}.$$

is known as Compton wavelength, where h is Planck's constant. For an electron, $\lambda_C = h/m_e c = 0.00243\,\text{nm}$ represents a "sub-nano" scale.

Compton's experimental results (Fig. 8.13) turned out fully consistent with Special Relativity.

Recently, by using inverse Compton scattering of electrons circulating in a ring accelerator in Grenoble (France), the velocity of light was tested along with a slight hypothetical violation of Lorentz symmetry. Considering that the direction of the laboratory rotates co-moving with the Earth once every 24 h, in 2010 the isotropy of the one-way light velocity and relativistic quantum electrodynamics (QED) was tested to an extraordinary accuracy of 10^{-14} (see Bocquet *et al.*, 2010).

8.5 Homework

1. Prove that, in a CM frame, the individual energies are related to Mandelstam variables via

$$E_1^* = \frac{s + m_1^2 - m_2^2}{2\sqrt{s}} \quad \text{and} \quad E_3^* = \frac{s + m_3^2 - m_4^2}{2\sqrt{s}}.$$

2. Prove that $\cos\vartheta = 1/\sqrt{\left(1 + \frac{2E_0}{E_1 - E_2}\right)\left(1 + \frac{2E_0}{E - E_0}\right)}$.
3. Show that Thales' circle results as a limiting case of the ellipse for momenta in relativistic collisions.
4. Verify $|\vec{p}_1|\,(E_1 + E_2) = \sqrt{(p_1^\mu\, p_{2\mu})^2 - m_1^2 m_2^2}$ in the case for which $\vec{p}_1 = -\vec{p}_2$, i.e. opposite momenta and $c = 1$.
5. Compton effect: Calculate the energy and momentum of a photon with a wavelength of 700 nm.
6. A photon with energy E_0 is scattered by a free electron initially at rest, in such a way that its scattering angle is the same as that of the dispersed photon ($\theta = \varphi$). Determine:

   (a) the angles θ and φ,
   (b) the energy and momentum of the scattered photon, and
   (c) the kinetic energy and momentum of the scattered electron.

7. A photon of 0.700 MeV is dispersed by a free electron in such a way that the scattering angle for the photon is twice the measured scattering angle for the free electron. Determine:

   (a) the scattering angle of the electron and
   (b) the final velocity of the electron.

8. Show that the magnitude of the velocity $v = |\vec{v}|$ of a particle having a de Broglie wavelength λ and Compton wavelength $\lambda_C = h/(mc)$ equals

$$v = c/\sqrt{1 + (\lambda/\lambda_C)^2}.$$

9. A photon with initial energy E_0 experiences Compton scattering in a certain angle due to a free electron of mass m_e, initially at rest. Using relativistic equations for the conservation of energy and momentum, derive

the following relationship for the final energy E' of the scattered photon:

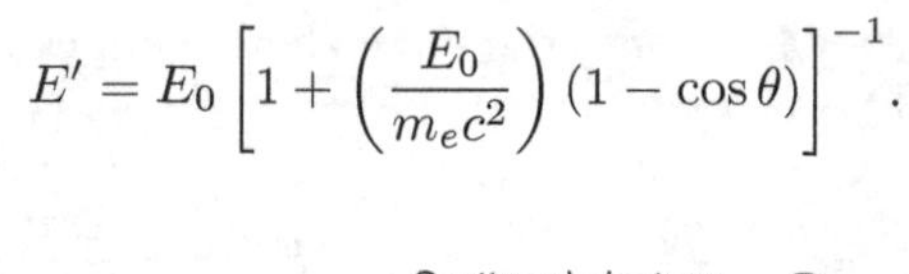

$$E' = E_0 \left[1 + \left(\frac{E_0}{m_e c^2} \right) (1 - \cos\theta) \right]^{-1}.$$

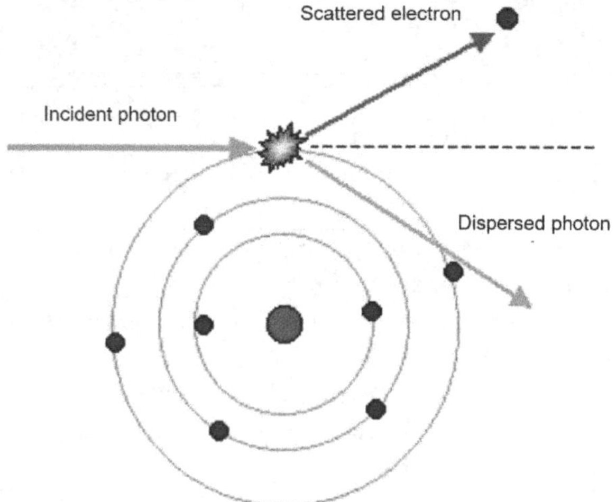

Reference

Berkovits, N. *et al.* (2017). *Memorial Volume for Stanley Mandelstam* (World Scientific, Singapore).

Chapter 9

Electrodynamics in 4D

The Lorentz invariance of the wave equation for the electric potential φ emerged before the formulation of Einstein's special relativity. The built-in covariance of Maxwell's equations comes to light more clearly when using four-vector or even tensor notation.

9.1 The Poincaré Transformations

In a Cartesian coordinate system, each event with its time and place is specified by the four coordinates:

$$x = (x^\mu) = \begin{pmatrix} x^0 \\ x^1 \\ x^2 \\ x^3 \end{pmatrix} = \begin{pmatrix} ct \\ \vec{r} \end{pmatrix}.$$

In the following, we consider two inertial systems, one primed I' and the other unprimed I, whose correspondence rule is a linear transformation that includes a translation x_0 of the spacetime origin:

$$x'^\mu = \Lambda^\mu_\nu x^\nu + x_0^\mu \longleftrightarrow x' = \Lambda x + x_0,$$

or infinitesimal,

$$dx'^\mu = \Lambda^\mu_\nu dx^\nu.$$

Here, Einstein summation for repeated indices is understood.

Now we are going to consider the following event, a spherical light front, which is emitted at time t_0 from $\vec{r}_0$ in I. In relation to I', the alike light wave was emitted at the same moment and has the coordinates $\vec{r}_0'$ for its center. Since in the two inertial frames the velocity of light is the same, the infinitesimal distance

$$ds'^2 = \eta_{\alpha\beta}dx'^{\alpha}dx'^{\beta} = \eta_{\alpha\beta}\Lambda^{\alpha}_{\mu}dx^{\mu}\Lambda^{\beta}_{\gamma}dx^{\gamma} = \Lambda^{\alpha}_{\mu}\eta_{\alpha\beta}\Lambda^{\beta}_{\gamma}dx^{\mu}dx^{\gamma} = ds^2$$

in Minkowski's spacetime remains invariant. A necessary and sufficient condition for the interval to be equal for the same event in two inertial frames of reference is

$$\Lambda^{\mathrm{T}}\eta\Lambda = \eta \longleftrightarrow \Lambda^{\alpha}_{\mu}\eta_{\alpha\beta}\Lambda^{\beta}_{\nu} = \Lambda^{\alpha}_{\mu}\Lambda_{\alpha\nu} = \eta_{\mu\nu}$$

which means that Λ are elements of a pseudo-orthogonal group $O(1,3)$ and T indicates the transposed matrix. The set of transformations:

$$x'^{\mu} = \Lambda^{\mu}_{\nu}x^{\nu} + x^{\mu}_{0} \quad \textit{and} \quad \Lambda^{\mathrm{T}}\eta\Lambda = \eta$$

form the Poincaré or inhomogeneous Lorentz group $\mathbf{P}$, with 10 parameters

$$\mathbf{P} = \{(\Lambda, x) \mid x \in R^4, \Lambda \in O(1,3), \Lambda^{\mathrm{T}}\eta\Lambda = \eta\}.$$

The group multiplication for the composition of two transformations is

$$(\Lambda_2, x_2) \circ (\Lambda_1, x_1) = (\Lambda_2\Lambda_1, \Lambda_2 x_1 + x_2).$$

This indicates that $\mathbf{P}$ is a semidirect product of translations and the Lorentz group, i.e. the translations x_0 also experience a Lorentz boost, see Appendix A for more details on the group structure.

Suppose that the inertial system I' moves in relation to the inertial reference system I with a relative adimensional velocity of $\vec{\beta} = \vec{v}/c$. The parallel $P_{\|}$ and perpendicular $P_{\perp}$ projectors in the direction of the relative velocity are

$$P_{\|} = \frac{1}{\beta^2}\vec{\beta}\cdot\vec{\beta}^{\mathrm{T}} \quad \text{and} \quad P_{\perp} = \mathbb{1} - P_{\|}$$

where $\beta = |\vec{\beta}|$ denotes its norm. Then the Lorentz boost can be written as a 4×4 matrix as follows:

$$\Lambda = (\Lambda^{\mu}_{\nu}) = \begin{pmatrix} \gamma & -\gamma\vec{\beta}^{\mathrm{T}} \\ -\gamma\vec{\beta} & P_{\perp} + \gamma P_{\|} \end{pmatrix}$$

with the Lorentz factor $\gamma = 1/\sqrt{1-\beta^2}$.

Given that $\Lambda^{\mathrm{T}}\eta\Lambda = \eta$ for pseudo-orthogonal groups, the matrices Λ have determinant ± 1, since

$$\det\,\Lambda^{\mathrm{T}}\det\,\Lambda = (\det\,\Lambda)^2 = 1 \rightarrow \det\,\Lambda = \pm 1.$$

The plus sign corresponds to the proper transformations including the complete reflection $\mathbb{PT}$, while the minus sign includes a parity reflection $\mathbb{P}$ or a inversion $\mathbb{T}$ of time. Together with charge conjugation $\mathbb{C}$, exact $\mathbb{CPT}$ invariance holds in quantum field theory (QFT).

9.2 Electric Currents in 4D

When including the charge density ρ, the 4D current density

$$j = (j^\mu) = \begin{pmatrix} c\rho \\ \vec{j} \end{pmatrix}, \quad \mu = 0, 1, 2, 3.$$

forms a four-vector (or Lorentz vector). This means that j^μ is transformed during a change of the inertial system as a vector x^μ and the coordinates of the inertial frames I and I' and the components j^μ and j'^μ of the 4D current density are applied in the following two corresponding systems. This specifies how the components of the Lorentz vector j^μ are transformed by a change of an inertial frame:

$$j'^\mu(x') = \Lambda^\mu_\nu j^\nu(x).$$

The continuity equation can now be re-written in a more elegant way as

$$\frac{\partial \rho}{\partial t} + \vec{\nabla} \cdot \vec{j} = \partial_\mu j^\mu(x) \equiv 0,$$

again using Einstein summation. Current conservation turns out to be compatible with Maxwell's equations.

By definition, the gradient operator in 4D is

$$\partial'_\mu \equiv \frac{\partial}{\partial x'^\mu} = \Lambda^\nu_\mu \partial_\nu.$$

For relative velocities $\vec{\beta} = v/c$, we can demonstrate the invariance of the continuity equation in any inertial system

$$\partial'_\mu j'^\mu(x') = (\Lambda^\nu_\mu \partial_\nu)(\Lambda^\mu_\alpha j^\alpha(x)) = \partial_\nu j^\nu(x) = 0.$$

The wave operator (d'Alembertian in Minkowski spacetime) has the form

$$\Box = \frac{\partial^2}{c^2\, \partial t^2} - \Delta = \partial_0^2 - \sum_{i=1}^{3} \partial_i^2 = \eta^{\mu\nu} \partial_\mu \partial_\nu = \partial^\mu \partial_\mu.$$

It is a differential operator which remains invariant under both, Lorentz or Poincaré transformations, namely

$$\Box = \partial^\mu \partial_\mu = \partial'^\mu \partial'_\mu = \Box'.$$

When d'Alembert formulated a wave equation for a vibrating string in 1747, this invariance may have been rudimentarily known.

9.3 Unifying Vector Potential

In electromagnetism, the vector potential $\vec{A}$ is a 3D vector field which allows to construct the magnetic field by derivation. The classic vector potential can be generalized to a four-vector by adding the electric potential φ as a time component. That is, toward a unification of the fields $\vec{E}$ and $\vec{B}$, the four-potential is given by

$$A^\mu = \begin{pmatrix} \varphi \\ \vec{A} \end{pmatrix} \Longrightarrow A'^\mu(x') = \Lambda^\mu_\nu A^\nu(x).$$

Using the Lorentz condition, i.e. $\partial_\mu A^\mu = 0$, the wave equation[1] can be written as

$$\Box A^\mu = \frac{4\pi}{c} j^\mu.$$

The Faraday tensor is defined as

$$F_{\mu\nu}(x) \equiv \partial_\mu A_\nu(x) - \partial_\nu A_\mu(x) = -F_{\nu\mu}(x).$$

And it is antisymmetric in its two indices. For the primed system we have the tensor transformation:

$$F'_{\mu\nu}(x') = \partial'_\mu A'_\nu(x') - \partial'_\nu A'_\mu(x') = \Lambda^\alpha_\mu \Lambda^\beta_\nu F_{\alpha\beta}(x).$$

Six of its components are linearly independent, which can be identified as follows:

$$F_{0i} = E_i \quad \text{and} \quad F_{ij} = -\epsilon_{ijk} B_k.$$

[1]Having Gauss as his teacher, as early as 1858 Bernhard Riemann (1867) proposed a relativistic invariant wave equation for the electromagnetic potential $\varphi = A^0$ in an attempt to reconcile, within a preliminary model of scalar electrodynamics, the Kohlrausch and Weber experiments of 1855. He correctly estimated the velocity of light in vacuum as $c \equiv 1/\sqrt{\varepsilon_0 \mu_0}$ from the values of the electromagnetic units then known. In 1886, Woldemar Voigt anticipated the invariance under, which now we call, rescaled Lorentz transformations of this wave equation. In the inhomogeneous Maxwell equation, the displacement current $\partial\vec{D}/\partial t$, where $\vec{D} = \varepsilon\vec{E}$, was anticipated in 1839 by James Mac Cullagh (1846), using the wave operator for the vector potential $\vec{A}$. Later, this turned out to be a key element in order to render electromagnetism relativistic invariant.

Accordingly, the Faraday tensor is given by

$$F_{\mu\nu} = \begin{pmatrix} 0 & E_1 & E_2 & E_3 \\ -E_1 & 0 & -B_3 & B_2 \\ -E_2 & B_3 & 0 & -B_1 \\ -E_3 & -B_2 & B_1 & 0 \end{pmatrix} = (\vec{E}, \vec{B}) \Longrightarrow (F^{\mu\nu}) = (-\vec{E}, \vec{B})$$

In Special Relativity, it is more convenient to use the Gaussian system of units, with $c = 1/\sqrt{\varepsilon_0\mu_0}$, as adopted here (Wipf, 2007).

Moreover, the field intensity tensor remains invariant under the gauge transformations:

$$A_\mu \longrightarrow A_\mu - \partial_\mu\theta.$$

In tensor notation, this gauge invariance follows from Poincaré's lemma

$$F_{\mu\nu} = \partial_\mu A_\nu - \partial_\nu A_\mu \longrightarrow \partial_\mu A_\nu - \partial_\nu A_\mu - (\partial_\mu\partial_\nu\theta - \partial_\nu\partial_\mu\theta) = F_{\mu\nu},$$

for functions $\theta(x)$ continuous in their second partial derivatives.

9.3.1 *Aharonov–Bohm effect*

Commonly, it was debated whether the vector potential $\vec{A}$ had a physical meaning or merely served as a mathematical device. Franz (1938) was the first to suggest that at least a phase change

$$\Theta = -\frac{e}{\hbar}\int \vec{A} \cdot d\vec{l}$$

in the wave function ψ of an electron can be observed. Let us suppose, as in Fig. 9.1, that two beams of an electron pass outside of a solenoid,

Figure 9.1: Vector potential $\vec{A}$ outside of a long solenoid.

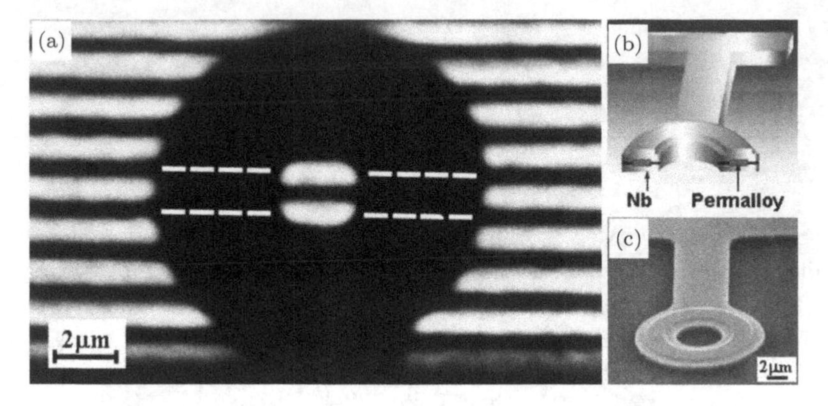

Figure 9.2: Shift in the interference pattern of the de Broglie matter waves of electrons. The permanent magnetic field of the Permalloy in the ring cannot be the cause, since it is shielded by the superconducting Niob cover.

one tangential to $\vec{A}$ and the other one in the opposite direction. The corresponding coherent electron waves undergo a phase change even though the magnetic field $\vec{B}$ is partially confined within the solenoid. This now called Aharonov-Bohm effect was convincingly confirmed by interference, see Batelaan & Tonomura (2009), after completely confining, via the Meissner effect, the magnetic field $\vec{B}$ into a superconducting torus.

9.3.2 *Electromagnetic fields derived from potentials*

The electric and magnetic potentials unified by the four-potential A^μ allow us to obtain the corresponding electromagnetic fields through

$$\vec{\boldsymbol{E}} = -\frac{\partial}{c\partial t}\vec{\boldsymbol{A}} - \vec{\nabla}\varphi \quad \text{and} \quad \vec{\boldsymbol{B}} = \vec{\nabla} \times \vec{\boldsymbol{A}},$$

or in components

$$\partial_0 A_i - \partial_i A_0 = E_i \quad \text{and} \quad \partial_i A_j - \partial_j A_i = \epsilon_{ijk} B_k.$$

For the relativistic formulation of Maxwell's equations in free space, we identify the inverse with the vector components

$$E_i = F_{0i} \quad \text{and} \quad B_i = -\frac{1}{2}\epsilon_{ijk} F_{jk}.$$

Now we are going to show that Maxwell's homogeneous equations

$$\nabla \cdot \vec{B} \equiv 0, \quad \nabla \times \vec{E} + \frac{1}{c}\frac{\partial \vec{B}}{\partial t} \equiv 0,$$

are identical to the Bianchi type covariant identity:

$$F_{[\mu\nu\rho]} \equiv F_{\mu\nu,\rho} + F_{\rho\mu,\nu} + F_{\nu\rho,\mu} = 0.$$

To verify the equations above, we consider all its components:

μ, ν, ρ	$F_{\mu\nu,\rho} + F_{\rho\mu,\nu} + F_{\nu\rho,\mu}$
0, i, 0	$F_{0i,0} + F_{00,i} + F_{i0,0} = 0$
i, j, 0	$F_{ij,0} + F_{0i,j} + F_{j0,i} = 0$
i, j, k	$F_{ij,k} + F_{ki,j} + F_{jk,i} = 0$

The first line vanishes identically, due to the anti-symmetry of the Faraday tensor. The second line is Maxwell's second homogeneous equation:

$$-\epsilon_{ijk}\partial_0 B_k + \partial_j E_i - \partial_i E_j = -\epsilon_{ijk}\left(\partial_0 B_k + \epsilon_{kpq}\partial_p E_q\right) = 0$$

or

$$-\left(\frac{1}{c}\frac{\partial \vec{B}}{\partial t} + \nabla \times \vec{E}\right) = 0,$$

where we have used the following identity:

$$\sum_k \epsilon_{ijk}\epsilon_{kpq} = \delta_{ip}\delta_{jq} - \delta_{iq}\delta_{jp}$$

for the 3D Levi-Civita completely antisymmetric tensor ϵ. The last equation in the table is identical to Maxwell's first homogeneous or Gauss equation

$$\partial_1 F_{23} + \partial_2 F_{31} + \partial_3 F_{12} = -\left(\partial_1 B_1 + \partial_2 B_2 + \partial_3 B_3\right) = -\nabla \cdot \vec{B} = 0.$$

For the non-homogeneous equations (initially in vacuum with isolated electric sources), we have

$$\nabla \cdot \vec{E} = 4\pi\rho, \quad \nabla \times \vec{B} - \frac{1}{c}\frac{\partial \vec{E}}{\partial t} = \frac{4\pi}{c}\vec{j}.$$

To rewrite it covariantly, we calculate the four-divergence of the Faraday tensor

$$\partial_\mu F^{\mu\nu} = \begin{pmatrix} \nabla \cdot \vec{E} \\ (-\partial_0 \vec{E} + \nabla \times \vec{B})^{\mathrm{T}} \end{pmatrix}$$

$$= (\partial_0, \partial_1, \partial_2, \partial_3) \begin{pmatrix} 0 & -E_1 & -E_2 & -E_3 \\ E_1 & 0 & -B_3 & B_2 \\ E_2 & B_3 & 0 & -B_1 \\ E_3 & -B_2 & B_1 & 0 \end{pmatrix}$$

and, using the definition of the four-vector $j^\mu = \{c\rho, \vec{j}\}$, we recover the electric 4D current as source:

$$\partial_\mu F^{\mu\nu} = \frac{4\pi}{c} j^\nu.$$

9.3.3 *Magnetic moments*

These Maxwell equations together with the relativistic Dirac equation

$$i\gamma^\mu \partial_\mu \psi = (mc/\hbar)\psi$$

constitute quantum electrodynamics (QED). The quantized (Dehghani, 2009) four-vector potential A_μ is minimally coupled via $\partial_\mu \to \partial_\mu + ieA_\mu$ in the Dirac equation. The source is the electron with a "radius" less than 2×10^{-20} m. Although almost pointlike (Dirac delta distribution), it has, as a Fermion, a magnetic moment $\mu = -(g/2)\mu_\mathrm{B}$ antiparallel to its intrinsic spin, due to its negative charge. This magnetic moment, in units of the Bohr magneton $\mu_\mathrm{B} = e\hbar/2m$, is measured with very high precision in a Penning trap. Compared to the calculations within QED, using higher order Feynman diagrams also including virtual photons, the agreement is astounding. See Gabrielse (2013) for more details.

9.4 Maxwell's Equations in Matter

In magnetic and/or polarizable media, we can define the *excitation* tensor as

$$H^{\mu\nu} = \begin{pmatrix} 0 & -c\,D_1 & -c\,D_2 & -c\,D_3 \\ c\,D_1 & 0 & -H_3 & H_2 \\ c\,D_2 & H_3 & 0 & -H_1 \\ c\,D_3 & -H_2 & H_1 & 0 \end{pmatrix}.$$

Then the non-homogeneous Maxwell's equations become

$$\partial_\mu H^{\mu\nu} = j^\nu.$$

For vanishing charge currents j, Maxwell's equations are invariant under a "duality rotation"

$$\star \begin{cases} \vec{H} \to \vec{E}, \\ \vec{D} \to -\vec{B} \end{cases}$$

known already to Von Laue.

In the old International System (SI) of Units, by the definition of the Ampere, the magnetic permeability is exactly $\mu_0 = 4\pi \times 10^{-7}$Newton/(Ampere)^2 and $\varepsilon_0 = 1/\mu_0 c^2$. In magnetic and/or polarizable media, the constitutive equations

$$\vec{D} = \varepsilon\, \vec{E}, \; \vec{B} = \mu\, \vec{H}$$

are involving via

$$\varepsilon = \varepsilon_0\, \varepsilon_r, \quad \mu = \mu_0\, \mu_r$$

the relative permittivity ε_r and permeability μ_r of the material. Thus, we can summarize Maxwell–Mac Cullagh's basic equations in the following table:

Non-homogeneous Maxwell equations	Constitutive relations	Homogeneous Maxwell equations
$\nabla \cdot \vec{D} = \rho$	$\vec{D} = \varepsilon\, \vec{E}$	$\nabla \times \vec{E} \equiv -\dfrac{\partial \vec{B}}{\partial t}$
$\nabla \times \vec{H} = j + \dfrac{\partial \vec{D}}{\partial t}$	$\vec{B} = \mu\, \vec{H}$	$\nabla \cdot \vec{B} \equiv 0$

Assuming linear, homogeneous, isotropic and non-dispersive charges in the material media, we can find a relationship between the vectors of electric intensity and magnetic induction through two parameters known as electric permittivity and the magnetic permeability:

$$\vec{D} = \varepsilon_0 \vec{E} + \vec{P},$$
$$\vec{B} = \mu_0 \big(\vec{H} + \vec{M}\big).$$

Here $\vec{P}$ is the polarization vector and $\vec{M}$ the magnetization of the material.

In general, their components are themselves (nonlinear) functionals of $\vec{E}$ and $\vec{B}$.

9.5 Electric and Magnetic Fields of a Charge in Uniform Motion

Consider a static elementary charge e at the origin of a reference frame I at rest. According to Maxwell's equations, the electric field simply corresponds to the radial component and the magnetic field components are identically null, that is, the electric and magnetic fields of an isolated charge are as follows:

$$\vec{E}(r) = \frac{e\vec{r}}{r^3} \quad \text{and} \quad \vec{B} = 0.$$

One wonders (with Einstein in 1905) what happens if we look at the same charge from another reference system I' that moves at a constant velocity v with respect to I. We would then experience an electric current that induces both electric and magnetic fields. The Lorentz transformations for these fields yield

$$\begin{aligned} E'_x &= E_x, & B'_x &= B_x \\ E'_y &= \gamma(E_y - \beta B_z), & B'_y &= \gamma(B_y + \beta E_z) \\ E'_z &= \gamma(E_z + \beta B_y), & B'_z &= \gamma(B_z - \beta E_y) \end{aligned}$$

Then, see Fig. 9.3, the electric fields are deformed and their enveloping radii r' form an oblate spheroid:

In order to get an expression for the fields in the primed system I', we must transform them. Taking into account that in the system I there is no magnetic field $\vec{B} = 0$, thus $B_x = B_y = B_z = 0$ initially, the resulting

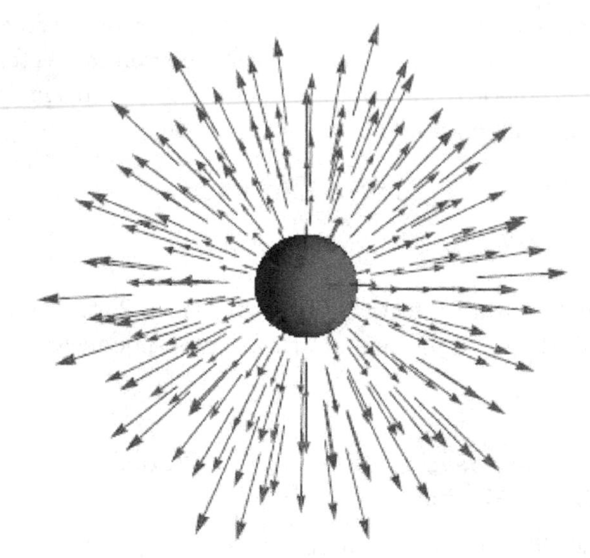

Figure 9.3: The electric field $\vec{E}$ of a point charge moving along the x axis, deformed by a Lorentz contraction to an oblate spheroid.

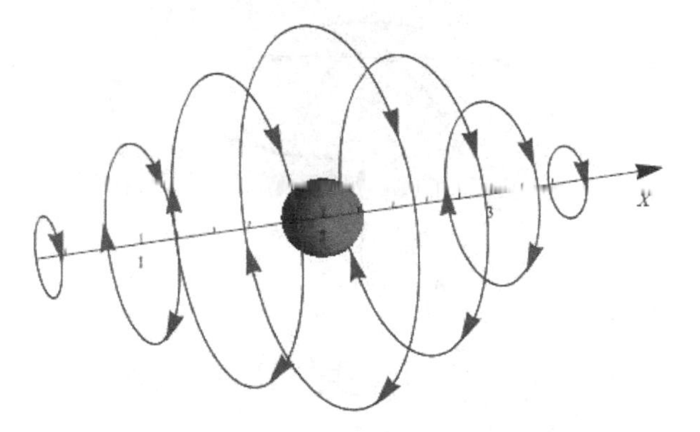

Figure 9.4: Magnetic field $\vec{B}$ of a point charge moving along the x axis.

electric and magnetic fields are

$$\vec{E}'(x) = \frac{e\gamma}{r'^{3/2}} \begin{pmatrix} x - \beta x^0 \\ y \\ z \end{pmatrix}, \quad \vec{B}'(x) = \frac{e\beta\gamma}{r'^{3/2}} \begin{pmatrix} 0 \\ -z \\ y \end{pmatrix}.$$

If the charge moves along the x axis, then the new enveloping radius satisfies $r'^2 = \gamma^2(x - \beta x^0)^2 + y^2 + z^2$, where $\beta = v/c$ and v is the constant relative velocity.

In the limit of non-relativistic velocities $v \ll c$, where $\gamma \backsimeq 1$, the simplified transformations

$$\vec{E'} \simeq \vec{E} + \vec{\beta} \times \vec{B}, \quad \vec{B'} \simeq \vec{B} - \vec{\beta} \times \vec{E}$$

would result for the electromagnetic fields in Gaussian units.

9.5.1 *Nanoscopically thin nickel needle as "fake" magnetic monopole*

The experimental search for magnetic monopoles has, so far, been in vain. The description of the monopole field by a local vector potential $A_\Phi = p(\pm 1 - \cos\vartheta)$, a "Dirac string", arises from coordinate 'patches' of a mathematical U(1) bundle (Mielke, 1986).

In order to avoid singularities of the vector potential $A_\Phi = p(1 - \cos\vartheta)$ at the axis of the infinitesimal solenoid (the Dirac string), cf. Fig. 9.5, it is sufficient to restrict its domain to $\vartheta \leq 3\pi/4$. On the other hand, for $\vartheta \geq \pi/4$, one can use the modified potential $A'_\Phi = -p(1 + \cos\vartheta)$. Thus,

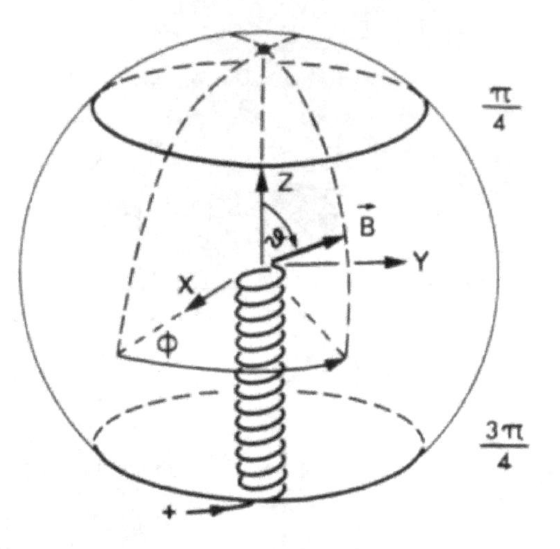

Figure 9.5: Coordinate patches for the Dirac monopole as a limit of an infinitesimal thin and long solenoid.

like for the sphere, at least two coordinate patches are needed for the gauge-dependent vector potential. In the overlapping region, the vector potentials are related by a gauge transformation. Nevertheless, the magnetic field turns out to be

$$\vec{B} = \frac{p}{r^3}\vec{r}$$

as expected for a monopole of pole strength p.

On the other hand, the wave function ψ of an electron should remain unique for a full pase shift $\Phi \to \Phi + \pi$ around the equator of the Dirac string. This uniqueness condition leads to the following famous charge quantization:

$$p = \frac{n\hbar c}{2e} = \frac{ne}{2\alpha} \simeq \frac{137}{2}ne,$$

where $\alpha = e^2/\hbar c$ is the fine structure constant of Sommerfeld.

Consequently, a real monopole would interact in a superstrong way with matter!

However, an effective monopole field can be produced at the tip of a long, nanoscopically thin magnetic needle, cf. Fig. 9.6. It was experimentally demonstrated (Béché *et al.*, 2014) that the interaction of this approximate magnetic monopole field in otherwise free space with a beam of electrons produces an electron vortex state, as theoretically predicted for a true magnetic monopole.

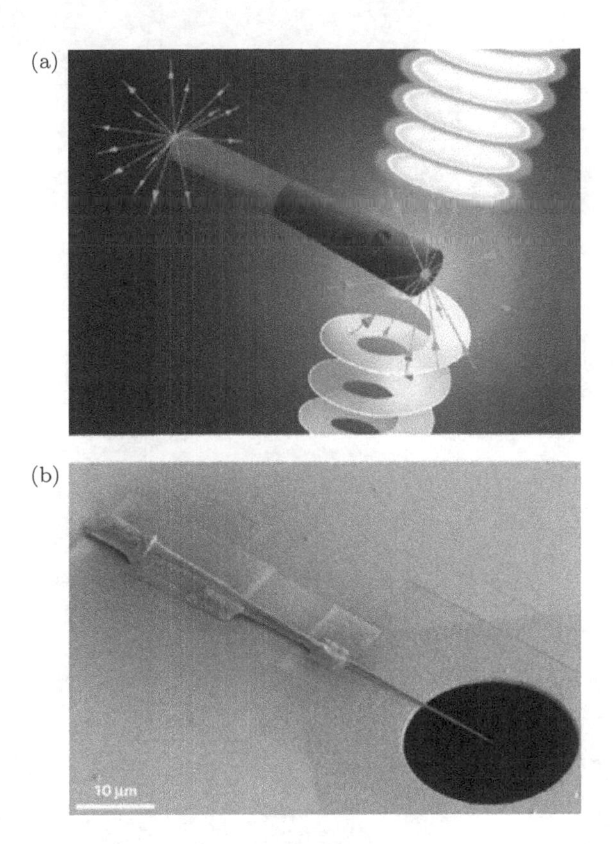

Figure 9.6: (a) Azimuthal Aharonov–Bohm phase shift of vortex type de Broglie waves of an electron. (b) Local monopole field from a thin Ni needle.

The Aharonov–Bohm effect can be used to understand the effects of such a monopole field on its surroundings, which is crucial for a possible observation and provides a better grasp of fundamental physics like QED.

9.6 Lorentz Force

The Lorentz force is the combination of electric and magnetic forces on a point charge or an electric current, see Fig. 9.7. For a particle subjected to an electric field combined with a magnetic field, the total electromagnetic force on that particle is given by the following Lorentz force:

$$\vec{F}_{\text{L}} = e\left(\vec{E} + \frac{\vec{v}}{c} \times \vec{B}\right).$$

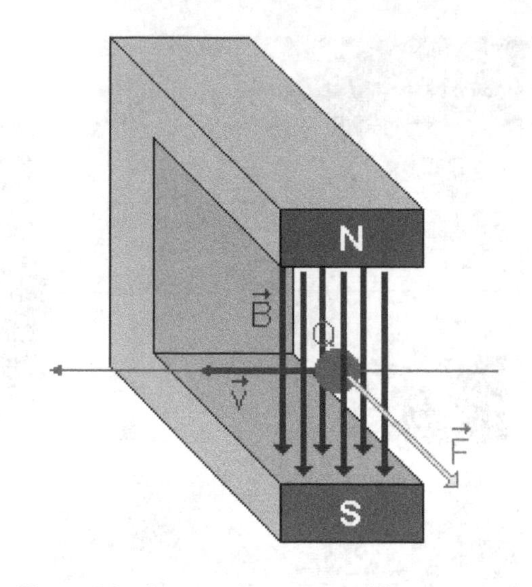

Figure 9.7: Lorentz force on an "electric swing".

Since

$$\begin{aligned}\vec{\beta}\cdot\vec{F}_{\rm L} &= e[\vec{\beta}\cdot\vec{E}+\vec{\beta}\cdot(\vec{\beta}\times\vec{B})]\\ &= e\vec{\beta}\cdot\vec{E},\end{aligned}$$

and

$$\vec{\beta}\cdot\vec{F} = m\gamma^3\vec{a}\cdot\vec{\beta}$$

hold for the scalar product, we can calculate the acceleration $\vec{\boldsymbol{a}}$ of a particle with mass m and charge e under the influence of the Lorentz force:

$$\begin{aligned}\vec{\boldsymbol{a}} &= \frac{1}{\gamma m}[\vec{F}_{\rm L}-\vec{\beta}(\vec{F}_{\rm L}\cdot\vec{\beta})]\\ &= \frac{1}{\gamma m}[\vec{F}_{\rm L}-e\vec{\beta}(\vec{E}\cdot\vec{\beta})]\\ &= \frac{e}{\gamma m}[\vec{E}-\vec{\beta}(\vec{E}\cdot\vec{\beta})+\vec{\beta}\times\vec{B}].\end{aligned}$$

There exist many applications of such a Lorentz force, for example, the mass spectrometer designed by F.W. Aston in 1919. The precision in the measurement of different atomic masses of isotopes has been improved with a velocity selector $\beta = E/B$, for perpendicular fields $\vec{E}$ and $\vec{B}$.

9.6.1 *Synchrotron*

The cyclotron was invented by E.O. Lawrence in 1932 to accelerate particles such as protons or deuterons to high kinetic energies. (Deuteron is the nucleus of deuterium, a stable isotope of heavy hydrogen, consisting of a proton and neutron strongly bound together.) The high energy particles can then be used to produce radioactive materials. For medical purposes, there exists the Betatron (cf. Fig. 9.8), which is used for the treatment of cancer. The operation of the cyclotron is based on the fact that the period $T = 2\pi m/e\,B$ of a charged particle moving relativistically inside a uniform magnetic field is independent of the velocity of the particle, cf. Fig. 9.9.

Nowadays, the cyclotron has been substituted by the synchrotron (Fig. 9.10) as a particle acelerator. Here, superconducting dipole magnets (in red) provide the Lorentz force in a direction perpendicular to both $\vec{v}$ and $\vec{B}$ ("inwards" as shown in the case above). Since the electron is moving with velocity $\vec{v}$, the force $\vec{F}_{\rm L}$ produces a centripetal acceleration causing a charged particle to move in (piecewise) circular orbits. In the ultra-high vacuum beam pipe, a *storage ring*, an electron can move around a closed loop consisting of curved (within the dipole magnets) and straight (outside the dipole magnets) sections.

Figure 9.8: Principle of a Betatron: An electron beam is moving in a circular path in a perpendicular quasi-homogeneous magnetic field sustained by Helmholtz coils. (The emission of light is caused by the excitation of some argon atoms in the bulb.)

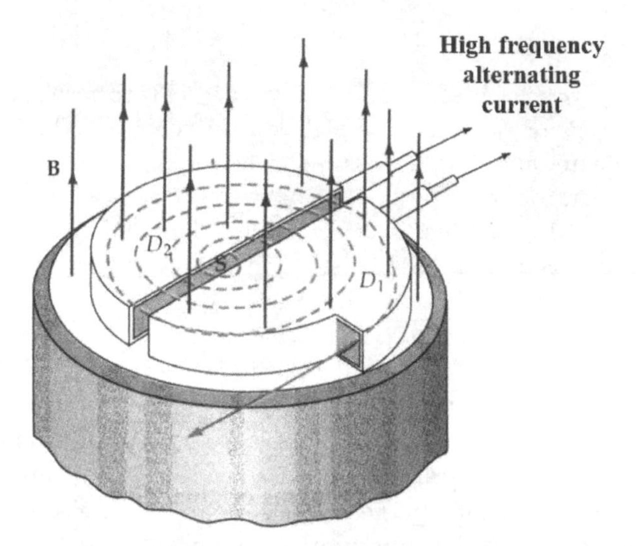

Figure 9.9: A cyclotron is a charged particle accelerator that combines the action of an alternating electric field, providing successive pulses, with a uniform magnetic field bending its trajectory and redirecting them again and again to the alternating electric field.

Figure 9.10: Synchrotron with injection Linacs and a wiggler magnet for the generation of synchrotron radiation.

Usually, one needs a source of energetic electrons to feed into the ring using a linear accelerator (Linac) which produces electrons at energies ranging from hundreds of MeV to several GeV. In some synchrotrons, a smaller "booster synchrotron", sited in between the Linac and the main synchrotron, is temporarily used during "start-up" (referred to as injection) just to bridge over the energy gap between the output-MeV of the Linac and the input-GeV required of the main synchrotron ring.

Wiggler (or undulatory) magnets induce some periodic acceleration which cause the electrons to generate coherent synchrotron radiation. This is employed for material research, or even biological tissues (Wille, 2013).

9.7 Homework

1. Show that in relativity, the electromagnetic wave equation does not vary, since the corresponding differential operator is invariant under Lorentz transformations. That is, in an adapted coordinate system, prove that

$$\Box \equiv \frac{1}{c^2}\frac{\partial^2}{\partial t^2} - \frac{\partial^2}{\partial x^2} - \frac{\partial^2}{\partial y^2} - \frac{\partial^2}{\partial z^2} = \frac{1}{c^2}\frac{\partial^2}{\partial t'^2} - \frac{\partial^2}{\partial x'^2} - \frac{\partial^2}{\partial y'^2} - \frac{\partial^2}{\partial z'^2} = \Box'.$$

2. Check the general transformations of the electric and magnetic fields

$$\vec{E}' = \gamma(\vec{E} + \vec{\beta}\times\vec{B}) - \frac{\gamma^2}{\gamma+1}\vec{\beta}(\vec{\beta}\cdot\vec{E}),$$

$$\vec{B}' = \gamma(\vec{B} - \vec{\beta}\times E) - \frac{\gamma^2}{\gamma+1}\vec{\beta}(\vec{\beta}\cdot\vec{B}).$$

applying a general Lorentz boost (Rindler, 1991). The last term arises from the parallel projector $(\gamma-1)P_{\|}$ when using the identity $\beta^2\gamma^2 \equiv (\gamma+1)(\gamma-1)$.

3. Derive the wave equation $\Box A^\mu = 0$ from Maxwell's equations in vacuum under the Lorentz condition $\partial_\mu A^\mu \cong 0$. (*Hint*: use the vector identity $\nabla\times\nabla\times\vec{A} = \nabla\times(\nabla\cdot\vec{A}) - (\nabla\cdot\nabla)\vec{A}$ for the triple cross product.)

4. Verify the previous equations for $\vec{v} = (v,0,0)^{\mathrm{T}}$ along the x axis. In the non-relativistic case, $v \ll c$ derive the Galilei-like transformations

$$\vec{E'} \simeq \vec{E} + \vec{\beta}\times\vec{B}, \quad \vec{B'} \simeq \vec{B} - \vec{\beta}\times\vec{E}.$$

Verify the invariance of the standard Maxwell Lagrangian proportional to $\vec{E}^2 - \vec{B}^2$ and the parity violating Pontryagin term $\vec{E}\cdot\vec{B}$ under Lorentz transformations. Check your results in the general case with a computer algebra software, such as Maple, Mathematica, or SageMath.

5. In a Betatron, a relativistic electron is kept in an orbit of radius R by the Lorentz force. Show (Ivanenko & Pomeranchuk, 1944) that the Lorentz factor is

$$\gamma \simeq \frac{e}{mc^2}\beta R|\vec{B}|.$$

References

Béché, A. *et al.* (2014). "Magnetic monopole field exposed by electrons", *Nat. Phys.* **10**, 26–29.

Dehghani, M. (2009). "Gupta–Bleuler quantization for massive and massless free vector fields", *Braz. J. Phys.* **39**(3), 559.

Gabrielse, G. (2013). "The standard models's greatest triumph", *Phys. Today,* **December**, 64–65.

Iwanenko, D. and Pomeranchuk, I. (1944). "On the maximal energy attainable in a Betatron", *Phys. Rev.* **65**, 343.

Mielke, E.W. (1986). "Magnetische Monopole in vereinheitlichten Eichtheorien", *Z. Naturforsch.* **41a**, 777–787.

Wille, K. (2013). "Synchrotron radiation", Joint Universities Accelerator School (JUAS) 28 January–1 February.

Chapter 10

Relativistic Ray Tracing

The intriguing question is: How do things look like when they move at relativistic velocities? This was solved thanks to the work of a colleague of Einstein, Anton Lampa, already in 1924, and later rediscovered by Penrose and Terrell, in 1959.

The shape, position, and orientation of an object moving at a velocity close to that of light can be quite different from what you would see at lower velocities. These optical deformations, known as aberrations, are reminiscent of images that occur in curved mirrors. They are due to two causes: angular compression and distortion. On the other hand, the color of the objects may vary due to the Doppler effect. Then, the intensity of light coming from objects moving at relativistic velocities changes also. So the image of the world that one would see would be affected by spatial contraction and time dilation. Aberrations, Doppler effect, and intensity changes have a common cause: The finite but huge velocity of light.

10.1 Relativistic Doppler Effect

The Doppler effect manifests itself in all kinds of waves, including sound and light. In the case of sound, we perceive this effect through changes in the frequency. For example, when a race car approaches us, the noise of its engine is heard more sharply and, when it moves away, the sound becomes more deep. Still, when a light-emitting object approaches us (or we approach it) the frequency of its light increases, and vice versa if it moves away. With light, frequency changes represent changes of color. An increased frequency of light produces a bluish object. A decrease in the frequency makes us see the object as more reddish. It has been observed that the spectral lines coming from distant stars or galaxies are somewhat more reddish than it should be if they were still. The redshift of light, according to Edwin Hubble's observations in 1920, allow us to deduce that the farthest galaxies are moving away from us at high velocity. These results are a proof of the expansion of the universe and its age of 13.8 billion years.

Now mathematically, let's consider the wave equation for the electric potencial φ in two frames of reference:

$$\Box\varphi = \Box'\varphi(t', x') = 0,$$

where

$$\Box = \frac{\partial^2}{c^2\,\partial t^2} - \vec{\nabla}\cdot\vec{\nabla}$$

is the operator of d'Alembert,[1] which is *invariant* under Lorentz transformations. For a plane wave, the solution is

$$\varphi = \mathrm{A}\exp i[\vec{k}\cdot\vec{x} - \omega t],$$

where $\vec{k}$ is the wave vector and ω is the angular frequency. Since the wave equation is invariant under general Lorentz transformations, so is the phase, that is,

$$-\eta_{\mu\nu}k^\mu x^\nu = \vec{k}\cdot\vec{x} - \omega t = \vec{k}'\cdot\vec{x}' - \omega' t'.$$

Consequently, $k^\mu = \{\omega/c, \vec{k}\}$ is transforming just like a four-vector $x^\nu = \{ct, \vec{x}\}$ of Lorentz, that is,

$$\omega' = \gamma(\omega - c\vec{\beta}\cdot\vec{k})$$

$$\vec{k}' = \vec{k} + \frac{\gamma - 1}{v^2}(\vec{v}\cdot\vec{k})\vec{v} - \gamma\vec{v}\,\frac{\omega}{c^2}.$$

Here $\gamma = 1/\sqrt{1-\beta^2}$ is the familiar Lorentz factor and $\vec{\beta} = \vec{\nu}/c$. The four-moment $p^\mu = \hbar k^\mu$ is obtained by multiplication with the reduced Planck constant $\hbar \equiv h/2\pi$.

By the wave equation, the (normal) dispersion relation

$$\vec{k}\cdot\vec{k} = \frac{\omega^2}{c^2}$$

---

[1] In 1747, Jean le Rond d'Alembert formulated and solved this differential equation for a vibrating string, essentially in 2D. Already in 1754 he proposed considering time as the fourth dimension, see Van Oss (1983) and Mayos *et al.* (2008).

is valid in vacuum and invariant under Lorentz transformations. Now suppose that the wave vector $\vec{k}$ forms an angle θ with the relative velocity. Then the scalar product $\vec{v} \cdot \vec{k} = vk\cos\theta$ can be simplified, where $k = |\vec{k}| = \omega/c$ is the norm of the wave vector. Thus,

$$\omega' = \gamma\omega\,(1 - \beta\cos\theta),$$

which is the *Doppler effect* for the angular frequency $\omega = 2\pi f$.

For a perpendicular wave vector, i.e. angle $\theta = 0$, we have

$$\omega' = \gamma\omega\,(1-\beta) = \frac{1-\beta}{\sqrt{(1-\beta)\,(1+\beta)}}\,\omega$$
$$= \sqrt{\frac{1-\beta}{1+\beta}}\,\omega = \omega/D$$

which is known as the *longitudinal* Doppler effect.

Here

$$D \equiv \sqrt{1+\beta}/\sqrt{1-\beta}$$

is the Doppler factor.

The transverse Doppler effect $\omega = \omega'\sqrt{1-\beta^2}$ can be more easily deduced from time dilation. At Einstein's suggestion, in 1938 Ives and Stilwell were the first to experimentally test this effect via a beam of excited light-emitting hydrogen atoms.

For the wavelength $\lambda = c/f = 2\pi c/\omega$, we obtain an inverse relationship as follows:

$$\lambda' = \sqrt{\frac{1+\beta}{1-\beta}}\,\lambda = \sqrt{\frac{c+v}{c-v}}\,\lambda,$$

where the Doppler factor $D = \sqrt{(c+v)/(c-v)}$ is independent of the wavelength. The relative velocity of objects in the cosmos can be inferred from the following relation:

$$1 + z = \frac{\lambda'}{\lambda} = 1 + \frac{\Delta\lambda}{\lambda}$$
$$= \sqrt{\frac{1+\beta}{1-\beta}} \simeq 1 + \beta.$$

where $z \equiv \Delta\lambda/\lambda$ corresponds to the redshift. There are quasars (bright objects called "quasi-stellar") so far away that they can have a redshift of $z > 3$.

Example: Distant galaxy

- The longest wavelength emitted by hydrogen in the Balmer series has a value of $\lambda_0 = 656\,\text{nm}$. In light from a distant galaxy, the value measured by spectroscopy is $\lambda' = 1458\,\text{nm}$. Find the velocity of remoteness or receding from that galaxy relative to Earth.

Solution: If we substitute these data in the formula for the Doppler factor squared, we have

$$\frac{1+\beta}{1-\beta} = \left(\frac{\lambda'}{\lambda_0}\right)^2 = \left(\frac{1458\,\text{nm}}{656\,\text{nm}}\right)^2 = 4.94$$

so that

$$1 + \beta = 4.94 - 4.94\beta$$

or

$$\beta = \frac{4.94 - 1}{4.94 + 1} = 0.663,$$

which is more than 66% of the velocity of light.

10.2 Aberration

To better understand the aberration of light, consider the following example: Suppose that someone stands in the rain and there is no wind. To keep from getting wet it is enough to put the umbrella over ones's head with the vertical cane. But if you start running and continue with the umbrella in an upright position, there will be drops of water that will reach you from

Figure 10.1: Tilting of rain in a moving frame.

the front. So, to avoid getting wet, the umbrella must be tilted in the direction of its movement. The angle of inclination in which you have to put the umbrella in order to avoid getting wet depends on the ratio of your velocity relative to that of the rain.

10.2.1 *Bradley discovered the aberration of light*

At the beginning of the 18th century, it was neither yet known how far the stars were, nor their proper motion. In an attempt to measure this distance, James Bradley (1693–1762) discovered in 1725 the phenomenon of the aberration of light, thereby unequivocally confirming the translational movement of the Earth and estimated the velocity of light. Bradley also discovered and measured the nutation or pitch of the Earth's poles.

Analogously, since the Earth moves and the light also (like the rain in the previous example), to observe a star, the telescope must be tilted a little in the direction of movement from the Earth. This inclination, which is necessary for the beam of light entering through the telescope's aperture to reach its bottom, is called light aberration. With his careful measurements, Bradley determined the velocity of light at 283 000 km/s, a value 5% less than the real one, but much more accurate than the one determined in 1676 by Rømer observing Jupiter's satellites.

If the wave vector $\vec{k}$ and $\vec{k}'$ has an angle θ or θ' with regard to the axes z or z' of propagation, we have

$$k'_x = k_x,\ k'_y = k_y,\ k'_z = \gamma\left(k_z - \frac{\beta}{c}\omega\right).$$

A polar coordinate system tells us that

$$\begin{aligned}\tan\theta' = \cot(\pi/2 - \theta') &= \frac{\sqrt{k_x'^2 + k_y'^2}}{k'_z} = \frac{\sqrt{k_x^2 + k_y^2}}{\gamma\,(k_z - \beta\omega/c)} \\ &= \frac{1}{\gamma}\frac{\sin\theta\ \omega/c}{(\cos\theta - \beta)\ \omega/c} = \frac{1}{\gamma}\frac{\sin\theta}{(\cos\theta - \beta)}.\end{aligned}$$

From the trigonometric identity $\tan(\alpha/2) = \sin\alpha/(1 + \cos\alpha)$, there results the equivalent formula

$$\tan\frac{\theta'}{2} = \sqrt{\frac{1+\beta}{1-\beta}}\tan\frac{\theta}{2} = D\ \tan\frac{\theta}{2},$$

where $D \equiv \sqrt{1+\beta}/\sqrt{1-\beta}$ is again the Doppler factor. This aberration formula is symmetric under an interchange of inertial frames.

10.2.2 *Superluminal velocities in twin jets?*

An apparently superluminal movement arises when an ejection of gas propagates at ultra-relativistic velocities so that it almost reaches the radiation it emits. This was first observed with a radio telescope (Mirabel & Rodríguez, 1998) with a wavelength of $\lambda = 35\,\text{cm}$ in twin microquasar jets. Let us consider a shiny jet knot located at the origin (center) at time $t = 0$. After a while, t has elapsed and the blob is away from the origin. The transverse displacement is $\beta\, ct\, \sin\theta$ (see Fig. 10.2).

However, since the knot is now closer to the observer by a distance $\beta\, ct\, \cos\theta$, the observer measures a shorter time $t' = \gamma(t - \beta\, ct\, \cos\theta)$ and therefore an apparent transverse adimensional velocity of

$$\beta_\perp \equiv \beta\, \sin\theta' = \frac{\beta\, \sin\theta}{\gamma\,(1 - \beta\cos\theta)}.$$

The Lorentz factor γ in this formula is important: Without it the transverse velocity would be superluminal, that is $\beta_\perp > 1$ when $\beta \simeq 1$ and θ is less than 90°. The **Galactic microquasar GRS 1915 + 105** was the first apparently superluminal source detected in our Milky Way. In Fig. 10.3, the radio images show a pair of jets ejected from the center. The brightest spot on the left has an apparent transverse velocity $\beta_\perp \simeq 1.25$. The true velocity of the knots is $\beta \simeq 0.92$, with $\theta \simeq 70°$. In the graph,

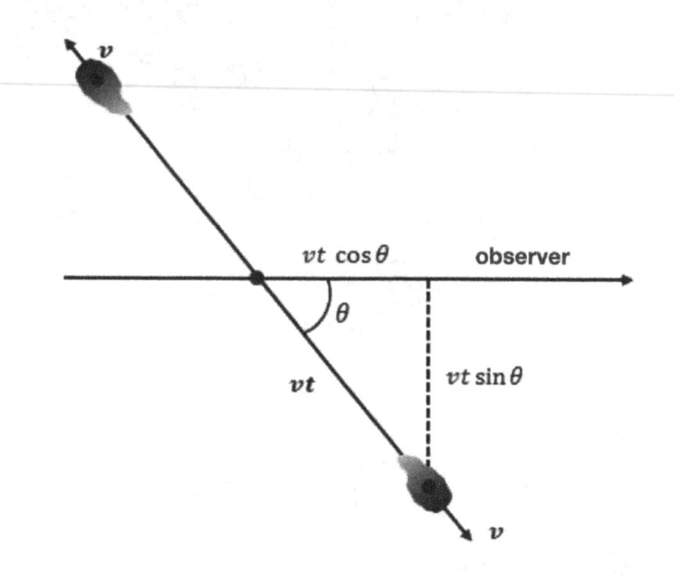

Figure 10.2: Geometry of twin jets.

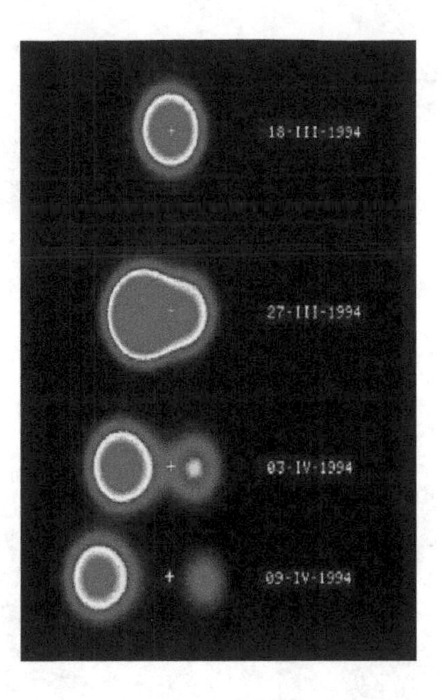

Figure 10.3: Jets in the microquasar GRS 1915 + 105 (Mirabel & Rodríguez, 1998).

AU is the astronomical unit, the average distance from Earth to the Sun, $1\,\mathrm{AU} \simeq 1.5 \times 10^8$km.

10.3 Invisibility of the Lorentz Contraction

Already in 1611, Christoph Scheiner used a "camera obscura" equipped with a telescope to safely project sunspots onto a screen (see Lefèvre, 2007). Modern adaptations of such "pin-hole cameras" are integrated in smartphones.

Let us make an idealized and instantaneous photo of an extended body by considering merely parallel rays: Then, all the photons of the same "face" of a cube arrive at the same time at the CCD (Charge-coupled device) of a camera, or the retina of an observer. The photons from the corner A must be emitted earlier to arrive at the same time (simultaneity!) at the retina. The additional distance caused by the relative velocity v is vl/c. Moreover, we know that for simultaneous events there is a Lorentz contraction $l \to l\sqrt{1-(v/c)^2}$, therefore, the BC side is shorter.

Then, a projection of a cube of the same size appears, but rotated with an angle α.

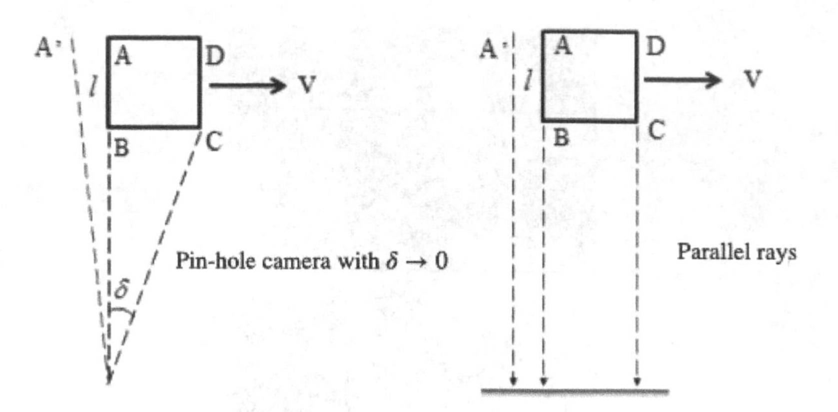

Figure 10.4: Relativistic photo of a cube.

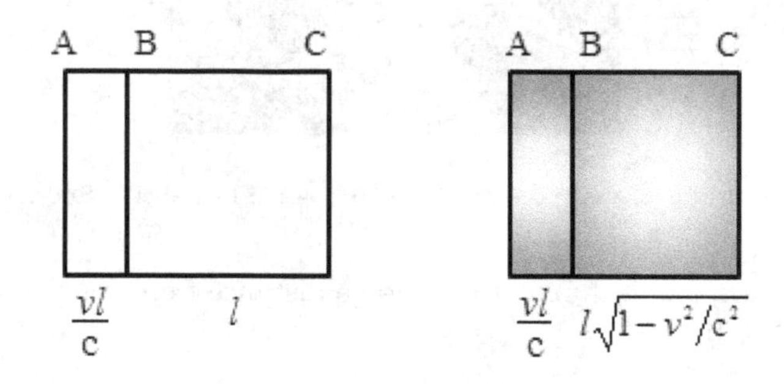

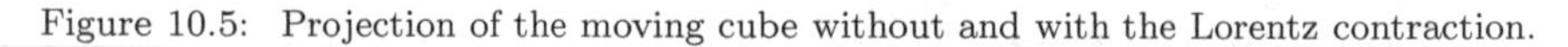

Figure 10.5: Projection of the moving cube without and with the Lorentz contraction.

If this geometrical interpretation is correct, the projection of the $\overline{AB}$ side should appear multiplied by

$$\sin\alpha = \frac{v}{c} = \beta$$

and the $\overline{BC}$ side by

$$\cos\alpha = \sqrt{1-\sin^2\alpha} = \sqrt{1-\frac{v^2}{c^2}} = \frac{1}{\gamma}.$$

We easily verify that the Pythagorean theorem is valid

$$\sin^2\alpha + \cos^2\alpha = \beta^2 + \left(\sqrt{1-\beta^2}\right)^2 = 1.$$

With such an idealized cube, we have explained Terrell's (1959) apparent "rotation" by a relativistic retardation effect. In the same year, Penrose

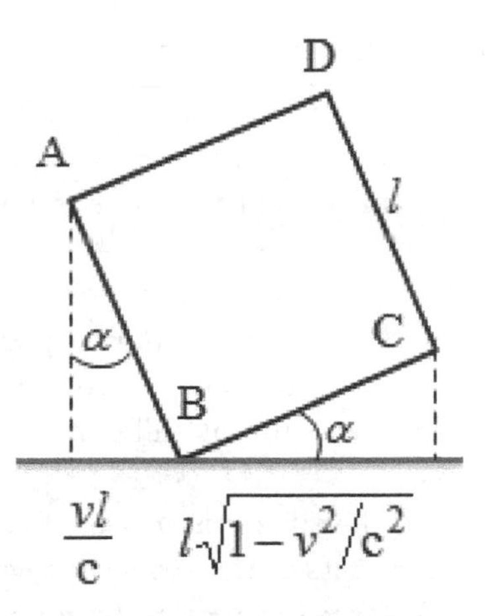

Figure 10.6: Apparent rotation of the cube by an angle α (Sexl & Urbandtke, 2001).

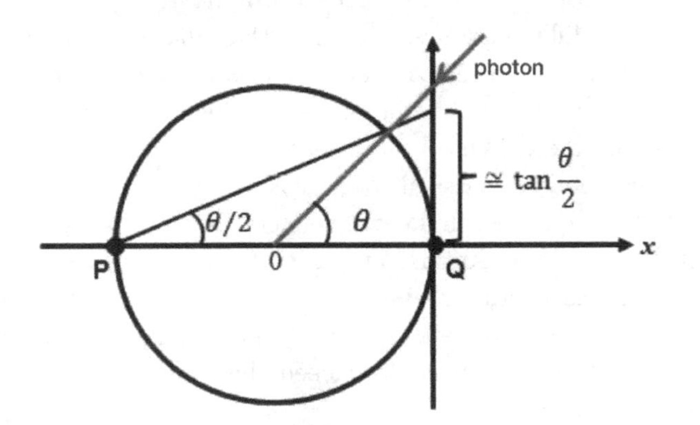

Figure 10.7: Stereographic projection to visualize the aberration geometrically.

realized that the aberration formula

$$\tan\frac{\theta'}{2} = D\,\tan\frac{\theta}{2}$$

has a geometric interpretation by stereographic projection: Suppose that the observer is in the center of his celestial sphere (Fig. 10.7).

A photon enters with an angle θ and its point of intersection with the sphere appears projected from P to a tangent plane at point Q. On this "screen", the shadow of a sphere is mapped to circles, as Penrose noted.

10.3.1 *Lampa–Terrell–Penrose rotation*

How does an extended object appear to move with relativistic velocity? Do you actually see a Lorentz contraction of an object, for example, a flat ellipsoid instead of a sphere? The answer is no, as illustrated in the following animation:

The movie MPEG4 of ≪spacetimetravel.org/tompkins/node1.html≫ shows the subsequent emission of light from a sphere compressed by the Lorentz contraction (for example, a nucleus in relativistic collisions of heavy ions). Although this light is emitted at different times, it must reach the observer at the same time.

Suppose the velocity of the sphere is $v = 0.95\ c$, its movement is from left to right, and at time $t = 0$, its center reaches the origin of the coordinate system. The light to be absorbed in motion by an opaque sphere is shown in yellow.

The observer sees at the same time the light that was emitted by the moving sphere at different times, i.e. when the sphere was in different positions. As a consequence, the observer sees a rotation and a slight distortion in the image of the sphere. In fact, some "backward" part of the sphere is visible to the observer. Thus, a picture, including the Lorentz contraction, is less distorted than its classical counterpart.

This is the Lampa–Terrell–Penrose effect: Except for a change of scale, it was first discovered by Anton Lampa in 1924, a colleague of Einstein.

Although a relativistic moving object appears rotated with an angle $\alpha = \arcsin \beta$, the emitted photons are also undergoing a redshift by a Doppler factor D. This shift could exceed the range of about 400–800 nm of the human color vision.

10.4 Visualization: Angular Compression and Increased Intensity

Imagine a person who is inside a train on a rainy and windless day. Raindrops, through the window, look vertical when the train is still and tilted when it is moving. Something similar happens with light. In Fig. 10.8, the central point represents a spaceship and the arrows are the rays of light (like the rain in the previous example) coming from all directions. In (a), the ship is at rest and in (b), the ship is moving at a relativistic velocity to the right. The light rays tilt in the direction of movement producing the effect of angular compression, in addition to the perspective, known already to Leonardo da Vinci.

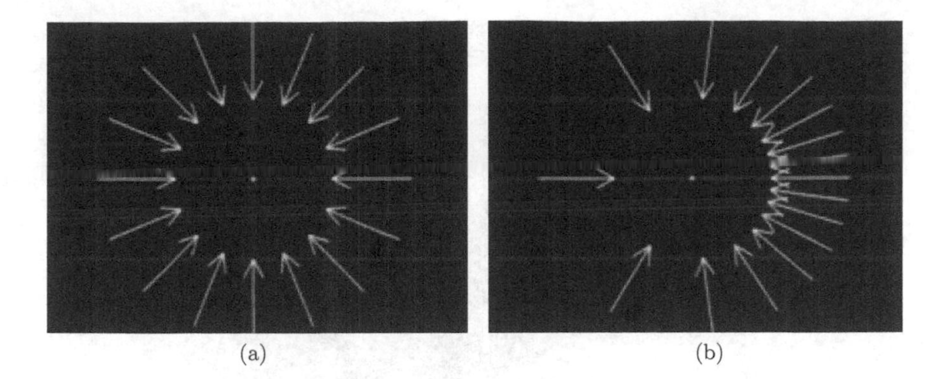

(a) (b)

Figure 10.8: Angular light compression effect.

The optical effect produced by this deflection of light rays is that objects in front appear compressed, while if you look back from the ship, everything will be enlarged.

Also, as seen in Fig. 10.8, angular compression tends to increase the amount of light coming from the front and reduce the amount coming from behind. From the perspective of someone at the center point of the figure, not only does a geometric deformation of objects occur that compresses forward and expand backward but also that objects appear brighter than those at the back.

This increase in intensity is proportional to the fifth power of the Doppler factor D, which may affect realistic visualizations in science-fiction movies.

In an animation (Kraus, 2008), one can clearly see the apparent rotation and distortion of a relativistic die (Fig. 10.9) and a streetcar (Fig. 10.10).

Mathematically, the pseudo-orthogonal Lorentz group $O(1,3) \cong SL(2, \mathbb{C})$ is locally isomorphic to the special (determinant = 1) linear transformations of the complex plane $\mathbb{C}$. Consequently, there is an equivalence of relativistic visualizations to Möbius types mappings, see Arnold & Rogness (2008).

Objects with relative velocities of $\beta = 0.9$ including the angular compression effect ("search light effect"), Doppler shift of the frequency, and intensity boost are displayed by "ray tracing" locally (see Müller *et al.* 2011, and Müller & Weiskopf, 2011, for electronic links and animations).

In scenarios where objects move in different directions, "ray tracing" in 4D is required, but its computational implementation is more sophisticated.

Figure 10.9: Relativistic appearance of a cube with velocity of $v = 0.9c$, seemingly rotated.

Figure 10.10: A "relativistic" streetcar appears rotated and distorted.

10.5 Ray Tracing in General Relativity

Ray tracing can also be generalized to General Relativity. Its objective is to compute images of astronomical bodies in the vicinity of compact objects, such as neutron stars, see Fig. 10.11, as well as the trajectories of massive particles in relativistic environments. This code is capable of integrating null and time-like geodesics not only for the Kerr metric of a rotating black hole (BH), but also in numerical solutions of General Relativity. Images and spectra have been simulated for a variety of astronomical targets, such as a moving star.

The first ray-tracing attempts in Einstein's General Relativity date back to the 1970s with works regarding the appearance of a star in orbit around a Kerr-type black hole, the derivation of a spectrum emitted from the accretion disk in terms of a transfer function, and the computation of the image of an accretion disk around a spherically symmetric Schwarzschild-type BH.

With the ray-tracing technique, you can visualize the image of a moving star, orbiting a Kerr black hole. The model is very simple, only the time-like geodesic of the star's center is calculated, and the star is defined as the points whose Euclidean distance to the center is less than a given radius R. An example is that of a black hole with a geometrically thin accretion disk revealing its "blue" horn due to the Doppler effect.

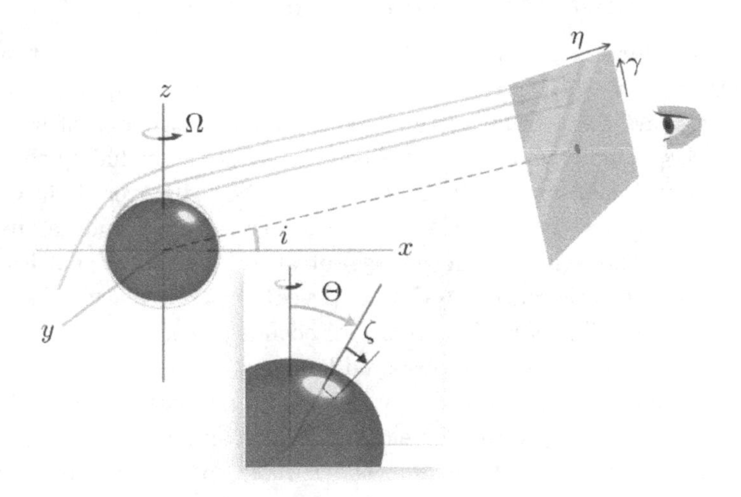

Figure 10.11: Principle of ray tracing at a long distance from a (rotating) neutron star (NS). Hence, geodesics become parallel upon arrival. Light-like or null geodesics are backwards-integrated to the observer, being located at an inclination i with respect to the equatorial plane, see Oliva & Frutos-Alfaro (2021).

Figure 10.12: Schwarzschild-type black hole (Vincent *et al.*, 2011).

There is a generally relativistic ray-tracing code (Vincent *et al.*, 2011) that reveals that the flux is proportional to the specific intensity emitted. The resulting image is shown in Fig. 10.12 for a Schwarzschild-type black hole and its accretion disk.

10.5.1 *Relativistic jet emanating from the galaxy M87*

The Event Horizon Telescope (EHT) had received radio waves from the elliptical galaxy M87 (number of the Messier catalogue), some 55 million light-years away from Earth. This giant holds trillions of stars and helps anchor roughly another two thousand galaxies — including the Milky Way — that make up the Virgo Cluster. It was discovered by Charles Messier all the way back in 1781. Three orders of magnitude as massive as the one at the center of our galaxy but also three orders as distant, its central active galatic nuclei (AGN) supposedly to be a black hole (M87*) has an apparent diameter of 40 microarcseconds (μas).

Its relativistic jet has been observed long ago by the Hubble space telescope. It has previously spotted a jet of high-energy particles from the center of M87, moving close to the velocity of light and extending more than a thousand light-years. Meanwhile, the Hubble space telescope observed a strange blob of matter in the jet, known as HST-1, which occasionally brightens and dims. Recently, the CHANDRA satellite measured via X-rays the proper relativistic motion of these knots (see Fig. 10.13).

When a disk rotates, spacetime gets deformed and the effect of frame dragging induces a helical magnetic field. Then, it is believed that Lorentz

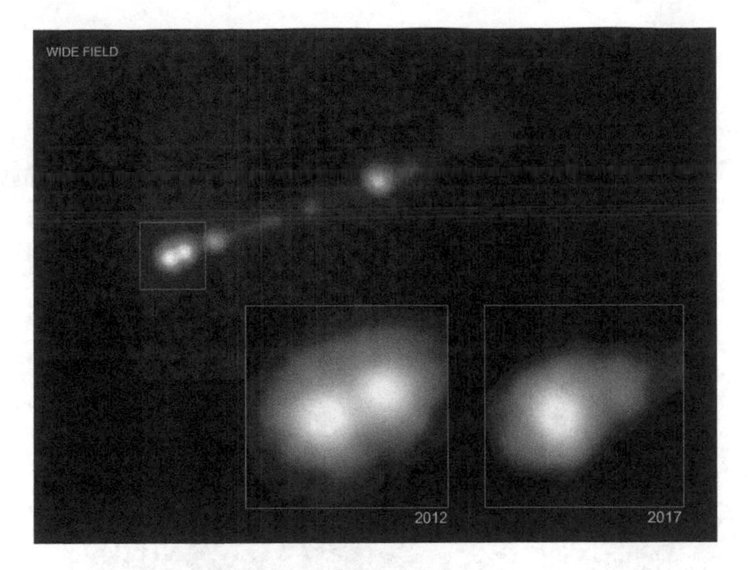

Figure 10.13: Powerful jets stream from the M87's center like a cosmic searchlight, extending out to at least 3 300 light-years. The X-ray jet originates from entangled magnetic fields in the accretion disk (Image credit: NASA/CXC/SAO/B. Snios *et al.*, 2019).

forces accelerate the plasma to form a relativistic jet perpendicular to the disk, similar to an electromagnetic levitation force acting on the conducting ring of E. Thompson.

The technique of linking distant radio telescopes to form a virtual telescope almost the size of the Earth, known as *very long baseline interferometry* (VLBI), is not new. None of the instruments are physically connected. Instead, they save data with exact time marks provided by local atomic clocks: That information was collated by supercomputers (Grant, 2019). Eight telescopes at radio wavelengths collaborated to view something like an accretion disk almost perpendicular to the line of sight.

Once the data were calibrated, the researchers had to turn them into one static image: The long-sought silhouette of a black hole. Beyond the event horizon, the huge forces involved will super-heat surrounding gases, as well as warping spacetime itself. The $42 \pm 3\mu$as diameter of the photon ring surrounding an orb of darkness, combined with the AGN's distance from Earth, allowed the team to estimate its mass: About 6.5 billion solar masses.

The brighter southern region is the result of the Doppler effect; the plasma of the accretion disk there is moving towards us (see Fig. 10.14). And deep within the dark central circle, with a radius no larger than 40% that of the visible plasma ring, would be the BH's event horizon.

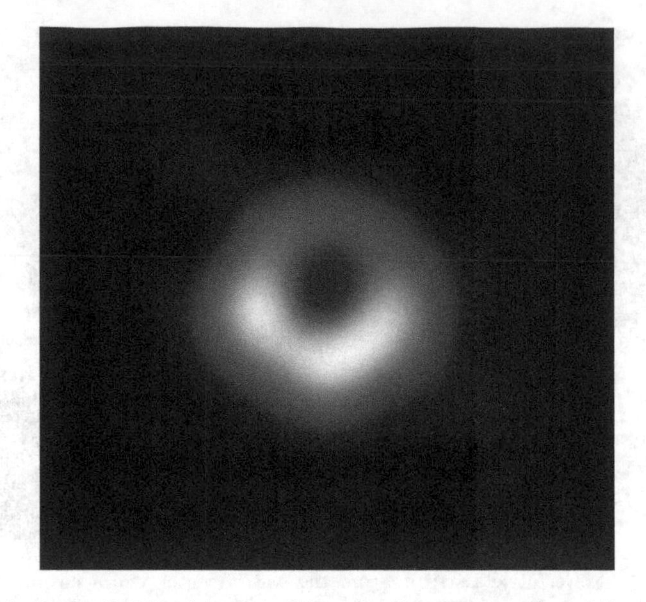

Figure 10.14: In April 2019, eight ground-based radio telescopes, forged through international collaboration, captured the first indirect evidence of the shadow of a supermassive black hole in the centre of M87 (Image credit: Event Horizon Telescope collaboration *et al.*).

10.6 Homework

1. Calculate the angle θ between the jet and the line of sight from the observed velocities $\beta_{\parallel} \simeq 6.3$ and $\beta_{\perp} \simeq 2.9$ for the HST-1 knot emanating from M87. Although apparently superluminal (Snios *et al.*, 2019), deduce from the transverse velocity $\beta_{\perp}$, including the Lorentz factor γ, that the true velocity is less than that of light, i.e. $\beta \lesssim 1$.
2. Stir-up (rotating !) vortices in water in order to let parallel light (e.g. from the sun) produce disk-like circular shadows.

   Take a photo of your "homemade black hole" and explain how caustics project a light-ring around your analogue BH, cf. Kiehn (2001).

References

Grant, A. (2019). "What it took to capture a black hole", *Phys. Today, Res. Technol.* 11 April 2019.

Kiehn, R.M. (2001). "Cosmic Strings in a Swimming Pool: Falaco Solitons", archiv: gr-qc/0101098.

Lefévre, W. (2007). *Inside the Camera Obscura: Optics and Art Under the Spell of the Projected Image* (Max-Planck Institute for the History of Science, Berlin).

Oliva, G.A. and F. Frutos-Alfaro (2021). "Effects of the treatment of the mass quadrupole moment on ray-tracing applications for rapidly rotating neutron stars", *Monthly Notices of the Royal Astronomical Society*, **505** Issue 2, August 2021, pp. 2870–2885.

Snios, B. *et al.* (2019). "Detection of superluminal motion in the X-Ray jet of M87", *Astrophys. J.* **879**, 8 July 1.

Appendix A

Spacetime Groups

A.1 Group Axioms

A group $(G, \circ)$ is a set G of elements $\{a, b, c, \ldots\}$ in which an internal composition law $\circ$ has been defined such that it satisfies the following axioms:

(i) Closure: $a \circ b = c \in G$.
(ii) Associativity: $a \circ (b \circ c) = (a \circ b) \circ c$.
(iii) Identity element: $e \circ a = a \circ e = a$
(iv) Inverse element: $a^{-1} \in G : a \circ a^{-1} = a^{-1} \circ a = e$

A.2 Rotations and Euler Angles

The Euler angles are of considerable importance in the study of the rotation of a rigid solid and can be defined through the following sequence of operations: rotating the coordinate system $\vec{x} \equiv (x, y, z)^{\mathrm{T}}$ around the z axis counterclockwise by an angle φ. The resulting coordinate system is $\vec{x}' = (x', y', z')^{\mathrm{T}}$. This system is then rotated counterclockwise by an angle θ around the x' axis, obtaining the coordinate system $\vec{x}'' = (x'', y'', z'')^{\mathrm{T}}$. If this system is then rotated around the z'' axis again counterclockwise, the final coordinate system $\vec{x}''' = (x''', y''', z''')^{\mathrm{T}}$ is obtained (see Fig. A.1).

In matrix form, the transformation rules corresponding to each operation are

$$\vec{x}' = \mathbb{R}(\varphi)\vec{x}, \quad \vec{x}'' = \mathbb{R}(\theta)\vec{x}', \quad \vec{x}''' = \mathbb{R}(\psi)\vec{x}''.$$

By consecutive application of these transformations

$$\vec{x}''' = \mathbb{R}(\psi)\mathbb{R}(\theta)\mathbb{R}(\varphi)\vec{x} = \mathbb{R}(\psi, \theta, \varphi)\vec{x}$$

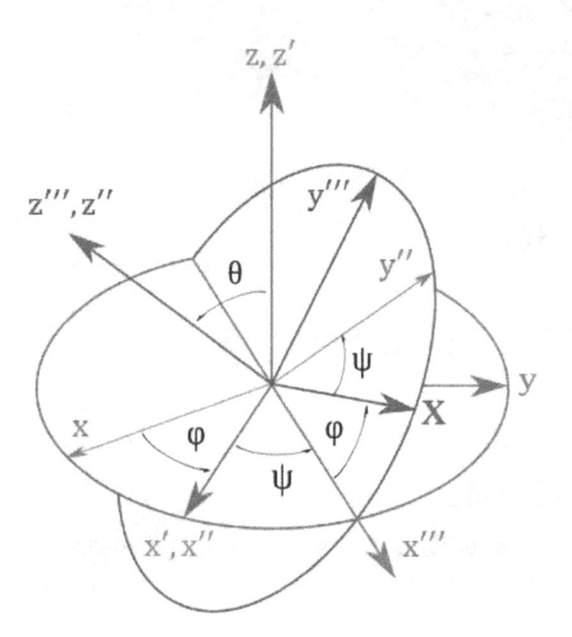

Figure A.1: Euler angles ψ, θ, φ.

is obtained. In the previous expression,

$$\mathbb{R}(\psi,\, \theta,\, \varphi\,) := \mathbb{R}(\psi)\mathbb{R}(\theta)\mathbb{R}(\varphi)$$

is the Euler matrix, where the subgroups are represented by the 3×3 matrices

$$\mathbb{R}(\varphi) = \begin{pmatrix} \cos\varphi & \sin\varphi & 0 \\ -\sin\varphi & \cos\varphi & 0 \\ 0 & 0 & 1 \end{pmatrix}, \quad \mathbb{R}(\theta) = \begin{pmatrix} 1 & 0 & 0 \\ 0 & \cos\theta & \sin\theta \\ 0 & -\sin\theta & \cos\theta \end{pmatrix},$$

$$\mathbb{R}(\psi) = \begin{pmatrix} \cos\psi & \sin\psi & 0 \\ -\sin\psi & \cos\psi & 0 \\ 0 & 0 & 1 \end{pmatrix}.$$

Consequently, considering the abbreviations $\mathrm{c}\psi \equiv \cos\psi$, $\mathrm{s}\psi \equiv \sin\psi$, etc., the Euler matrix takes the rather involved following form:

$$\mathbb{R}(\psi,\, \theta,\, \varphi\,) = \begin{pmatrix} \mathrm{c}\psi\mathrm{c}\varphi - \mathrm{s}\psi\mathrm{c}\theta\mathrm{s}\varphi & \mathrm{c}\psi\mathrm{s}\varphi + \mathrm{s}\psi\mathrm{c}\theta\mathrm{c}\varphi & \mathrm{s}\psi\mathrm{s}\theta \\ -\mathrm{s}\psi\mathrm{c}\varphi - \mathrm{c}\psi\mathrm{c}\theta\mathrm{s}\varphi & -\mathrm{s}\psi\mathrm{s}\varphi + \mathrm{c}\psi\mathrm{c}\theta\mathrm{c}\varphi & \mathrm{c}\psi\mathrm{s}\theta \\ \mathrm{s}\theta\mathrm{s}\varphi & -\mathrm{s}\theta\mathrm{c}\varphi & \mathrm{c}\theta \end{pmatrix}.$$

A.3 Homework

1. Show explicitly that $\mathbb{R}^{-1} = \mathbb{R}^{\mathrm{T}}$ holds for orthogonal groups, where T denotes the transposed matrix (Goldstein, 1980).

A.4 Inhomogeneous Galilean Group

For Galilean transformations, it is convenient to consider a degenerate representation generated by quadratic matrices of 5×5 (unphysical) dimensions:

$$g(\mathbb{R}, \vec{v}, \vec{x_0}, t_0) = \begin{pmatrix} 1 & 0 & t_0 \\ \vec{v} & \mathbb{R} & \vec{x_0} \\ 0 & 0 & 1 \end{pmatrix}.$$

Here g is a function of 10 parameters and $\mathbb{R} \in SO(3)$ is an element of the orthogonal 3D subgroup of rotations.

Transitive action is given by matrix multiplication and results in a (degenerate) vector in five dimensions:

$$\begin{pmatrix} t' \\ \vec{x}' \\ 1 \end{pmatrix} = \begin{pmatrix} 1 & 0 & t_0 \\ \vec{v} & \mathbb{R} & \vec{x_0} \\ 0 & 0 & 1 \end{pmatrix} \begin{pmatrix} t \\ \vec{x} \\ 1 \end{pmatrix} = \begin{pmatrix} t + t_0 \\ \mathbb{R}\vec{x} + \vec{v}t + \vec{x_0} \\ 1 \end{pmatrix}.$$

The action includes, as seen, four translations, one in time by t_0 and three in space by $\vec{x_0}$.

The law of group multiplication explicitly leads to

$$g_2 \circ g_1 = \begin{pmatrix} 1 & 0 & t_2 \\ \vec{v_2} & \mathbb{R}_2 & \vec{x_2} \\ 0 & 0 & 1 \end{pmatrix} \begin{pmatrix} 1 & 0 & t_1 \\ \vec{v_1} & \mathbb{R}_1 & \vec{x_1} \\ 0 & 0 & 1 \end{pmatrix}$$

$$= \begin{pmatrix} 1 & 0 & t_1 + t_2 \\ \mathbb{R}_2\vec{v_1} + \vec{v_2} & \mathbb{R}_2\mathbb{R}_1 & \mathbb{R}_2\vec{x_1} + \vec{v}_2 t_1 + \vec{x}_2 \\ 0 & 0 & 1 \end{pmatrix},$$

revealing how the translations are also affected by rotations. Let us anotate the inverse transformation as

$$g^{-1} = \begin{pmatrix} 1 & 0 & -t_0 \\ -\mathbb{R}^{-1}\vec{v} & \mathbb{R}^{-1} & \mathbb{R}^{-1}\vec{v}t_0 - \mathbb{R}^{-1}\vec{x_0} \\ 0 & 0 & 1 \end{pmatrix}$$

A.5 Homework

1. Prove explicitly that $g \circ g^{-1} = \mathbb{1}$.

A.6 Poincaré Transformations

In Special Relativity, Galilean transformations are replaced by

$$x^{\mu} \to x'^{\mu} = \Lambda^{\mu}_{\nu}\left(\mathbb{R}, \vec{v}\right) x^{\nu} + x_0^{\nu}$$

which form a subgroup of the affine transformations.

As in the Galilei group, we have 10 parameters, the first sum has six parameters, while the second (the translations) have 4. Its representation via 5×5 matrices is

$$p\left(\Lambda, x_0\right) = \begin{pmatrix} \Lambda & x_0 \\ 0 & 1 \end{pmatrix} \in \mathbf{P}$$

where $\mathbf{P}$ is the Poincaré group and $x = x^{\mu}$ is a 4D vector.

The transitive action again on a (degenerate) five-vector is

$$\begin{pmatrix} \Lambda & x_0 \\ 0 & 1 \end{pmatrix} \begin{pmatrix} x \\ 1 \end{pmatrix} = \begin{pmatrix} \Lambda x + x_0 \\ 1 \end{pmatrix},$$

as expected.

Here the Lorentz group is a subgroup:

$$\tilde{\mathbf{\Lambda}} = p\left(\Lambda, 0\right) = \begin{pmatrix} \Lambda & 0 \\ 0 & 1 \end{pmatrix}$$

with exactly six parameters. The translations "live" in the coset space $\mathbf{P}/\Lambda$. The Lorentz group is non-singular, i.e. its determinant is non-zero.

A.7 Lorentz Group

Let's suppose that

$$\Lambda\left(\,\mathbb{R}, \vec{v}\,\right) = \begin{pmatrix} a(v) & b(v)\ \vec{v}^{\mathrm{T}}/v_{\max}^2 \\ -\gamma(v)\vec{v} & \mathbb{R}(\vec{v}) \end{pmatrix}$$

is its matrix representation, where $\vec{v}^{\mathrm{T}} = (v_x, v_y, v_z)$ is the transpose of the vector $\vec{v}$ and $v_{\max}$ is a maximum velocity specified afterwards.

Due to the isotropy of space, rotations can be rewritten as

$$\mathbb{R}(v) = c(v)\mathbb{1} + d(v)\frac{\vec{v}}{v} \otimes \frac{\vec{v}^{\mathrm{T}}}{v}$$

where $v = |\vec{v}|$ is the absolute value of the relative velocity.

The transitive action applied to a spacetime four-vector $\{t, \vec{x}\}$ is

$$\begin{pmatrix} t' \\ \vec{x}' \end{pmatrix} = \begin{pmatrix} a(v) & b(v)\vec{v}^{\mathrm{T}}/v_{\max}^2 \\ -\gamma(v)\vec{v} & \mathbb{R}(\vec{v}) \end{pmatrix} \begin{pmatrix} t \\ \vec{x} \end{pmatrix}$$

$$= \begin{pmatrix} a(v)t + b(v)\vec{v} \cdot \vec{x}/v_{\max}^2 \\ c(v)\vec{x} + \dfrac{d(v)}{v^2}(\vec{v} \cdot \vec{x})\,\vec{v} - \gamma(v)\vec{v}t \end{pmatrix}$$

being linear in $\vec{x}$ and in t. Since the rotation group $O(3)$ does not experience symmetry-breaking in vacuum, the transformations have the assumed isotropy with respect to $\vec{x}$. Consequently, the coefficients $a, b, c, d,$ and γ depend only on the absolute value $v = |\vec{v}|$ of the relative velocity.

A.7.1 *Lorentz group parameters*

Let us suppose that the **velocity of the origin**

$$\mathcal{O}' = \{\vec{t}' = 0,\ \vec{x}' = 0\},$$

of a reference frame I' of a traveler relative to I is $\vec{v} = \vec{x}/t$. This leads to the condition $0 = \{c(v) + d(v) - \gamma(v)\}\vec{x}$.

If $\mathbb{P}$: $\vec{x} \to -\vec{x}$ denotes a parity reflection in 3D, the transposed Λ^{T} of the Lorentz group satisfies $\Lambda^{\mathrm{T}}\eta\,\Lambda = \eta$. This is equivalent to $\Lambda^{-1} = \eta^{-1}\Lambda^{\mathrm{T}}\eta$ due to the parity representation $\mathbb{P} : x = \eta x$ for four-dimensional vectors $x = \{t, \vec{x}^{\mathrm{T}}\}$, where $\eta \equiv \mathrm{diag}(1, -1, -1, -1)$ is the Minkowski metric.

Then the inverse element is such that

$$\Lambda^{-1}(\mathbb{R}, \vec{v}) = \Lambda^{\mathrm{T}}(\mathbb{R}, -\vec{v})$$

for pseudo-orthogonal groups. Physically, the inverse element simply corresponds to the replacement of $\vec{v}$ for $-\vec{v}$ in the direction of the velocity and inertial frames $I \to I'$. Then group theory leads us to the matrix condition:

$$\Lambda^{-1} \circ \Lambda = \left(\begin{array}{c|c} a & \gamma\,\vec{v}^{\mathrm{T}}/v_{\max}^2 \\ \hline -\,b\vec{v} & \mathbb{R}^{\mathrm{T}}(-\vec{v}) \end{array}\right) \left(\begin{array}{c|c} a & b\,\vec{v}^{\mathrm{T}}/v_{\max}^2 \\ \hline -\,\gamma\vec{v} & \mathbb{R}(\vec{v}) \end{array}\right)$$

$$= \left(\begin{array}{c|c} a^2 - \gamma^2\, v^2/v_{\max}^2 & (ab + \gamma c + \gamma d)\,\vec{v}^{\mathrm{T}}/v_{\max}^2 \\ \hline -\,(ba + c\gamma + d\gamma)\vec{v} & c^2 + \left(2cd + d^2 - b^2\dfrac{v^2}{v_{\max}^2}\right)\dfrac{\vec{v}}{v} \otimes \dfrac{\vec{v}^{\mathrm{T}}}{v} \end{array}\right)$$

$$= \begin{pmatrix} 1 & 0 \\ 0 & \mathbb{1} \end{pmatrix}.$$

A.7.2 *Algebraic conditions*

From the above, we arrive at five conditions for the five arbitrary coefficients of the Lorentz transformations:

$$\begin{aligned} & c + d - \gamma = 0 \\ & a^2 - \gamma^2 \frac{v^2}{v_{\max}^2} = 1 \\ & ab = -\gamma(c + d) \\ & c^2 = 1 \\ & d(2c + d) = b^2 \frac{v^2}{v_{\max}^2}. \end{aligned}$$

Their algebraic solution is

$$\begin{aligned} & c = 1 \quad (c = -1 \text{ would correspond to the parity reflection } \mathbb{P}) \\ & d = \gamma - 1, \quad \to \gamma^2 - 1 = b^2\, v^2/v_{\max}^2 \\ & ab = -\gamma^2, \quad a^2 = 1 + \gamma^2\, v^2/v_{\max}^2 \end{aligned}$$

Taking the second-last equation above and multiplying itself by a^2, we find:

$$\begin{aligned} a^2 b^2\, v^2/v_{\max}^2 &= \gamma^4\, v^2/v_{\max}^2 = \left(1 + \gamma^2\, v^2/v_{\max}^2\right)\left(\gamma^2 - 1\right) \\ &= \gamma^4\, v^2/v_{\max}^2 + \left(1 - v^2/v_{\max}^2\right)\gamma^2 - 1. \end{aligned}$$

Since the fourth power in γ cancels in the last relation, we arrive at the:

A.7.3 *Lorentz factor*

$$\boxed{\gamma(\vec{v}) \equiv \frac{1}{\sqrt{1 - \vec{v}^2/v_{\text{max}}^2}}}$$

Additionally, $a = \pm\gamma = -b$. (The negative sign would correspond to the reflection $\mathbb{T}$ of time.)

A.8 Homogeneous and Isotropic Spacetime Transformations

From the latter, we get:

$$t' = \gamma\left(t - \frac{\vec{v}\cdot\vec{x}}{v_{\text{max}}^2}\right)$$

$$\vec{x}' = \vec{x} + \frac{\gamma - 1}{v^2}\left(\vec{v}\cdot\vec{x}\right)\vec{v} - \gamma\vec{v}t,$$

cf. Rindler (1991) or Sexl & Urbantke (2001).

Special cases:

(a) Special Galilean transformations. When $v_{\text{max}} \to \infty$, which implies that $\gamma \to 1$, we have

$$t' = t$$

$$\vec{x}' = \vec{x} - \vec{v}t$$

(b) Lorentz transformations result when putting $v_{\text{max}} = c$, the velocity of light in vacuum.

Reference

Goldstein, H. (1980). *Classical Mechanics*, 2nd ed. (Addison-Wesley).

Appendix B

Units of the International System

Time [s]	The **second** is the duration of 9 192 631 770 periods of radiation corresponding to the transition between the two hyperfine levels of the ground state of Caesium 133.
Length [m]	The **meter** is the path length traveled in vacuum by light for a time of 1/299 792 458 seconds, i.e. $1\,\mathrm{m} = (c/299\ 792\ 458)\ \mathrm{s}$
Mass [kg]	The **kilogram** is equal to the mass of the international prototype. 

Electric current [A]	The **ampere** is the intensity of constant electric current that is maintained in two conductors and would produce a force equal to 2.0×10^{-7} Newton per meter in length. These conductors must be parallel, straight, almost infinite length, with negligible circular section and located at a distance of one meter from each other in vacuum.
Temperature [K]	The **kelvin** is the unity of the thermodynamical temperature, defined by the fraction 1/273.16 of the absolute temperature of the triple point of water.
Luminous intensity [cd]	The **candela** is the luminous unit in a given direction of a source that emits a monochromatic frequency radiation of 540×10^{12} Hertz and whose radiant intensity in that direction is 1/683 W per steradian.
Amount of substance [mol]	The **mole** is the amount of substance in a system that contains as many elemental entities as there are atoms in 0.012 kilograms of the carbon isotope C_{12}. The elemental entities must be specified, since it can be atoms, molecules, ions, electrons, or other types of particles or groups of particles.

On World Meteorology Day 2019, new definitions of the fundamental units of the International System of Units came into effect:

Time [s]	The **second** is based on the transition frequency $\Delta f_{Cs} = 9\ 192\ 631\ 770$ Hz of radiation corresponding to the transition between the two hyperfine levels of the ground state of Caesium 133.
Length [m]	The **meter** is based on the fixed velocity $c = 299\ 792\ 458$ m/s of light in vacuum, i.e. $1\,\mathrm{m} = (c/299\ 792\ 458)$ s.
Mass [kg]	The **kilogram** is defined in terms of the fixed numerical value of the Planck constant $h = 6.62607015 \times 10^{-34}$ J s, the latter unit being equal to kg m^2 s^{-1}.
Electric current [A]	The **ampere** is the unit of electric current. It is defined by taking the fixed numerical value of the elementary charge e to be $1.602176634 \times 10^{-19}$ C when expressed in the unit Coulomb C, equal to A s, where the second is defined in terms of Δf_{Cs}.
Temperature [K]	The **kelvin** is defined by taking the fixed numerical value of the Boltzmann constant k to be $1.380\ 649\ \times\ 10^{-23}$J/K, the latter unit being kg m^2 s^{-2} K^{-1}.
Luminous intensity [cd]	The **candela** is the unit of luminous intensity of a source that emits a monochromatic radiation frequency of 540×10^{12} Hz and whose radiant intensity in that direction is 1/683 W per steradian, the latter being equal to cd sr kg^{-1} m^{-2} s^{3}.

Amount of substance [mol]	The **mole** is the amount of substance in a system that contains as many elemental entities (atoms, molecules, ions, electrons, or other particles or precise groups of particles) as the fixed Avogadro constant: $N_A = 6.022\ 140\ 76 \times 10^{23}/\text{mol}$  In the defunct International Avogadro project, the number of atoms in an ideal silicon sphere would be counted via X-ray crystallography and optical interferometry.

(see BIPM 2019 for more details.)

Derived units in the SI

Magnitude	**Name**	**Symbol**	**Basic units**	**Expression in terms of other units**
Plane angle	radian	rad	m/m	
Frequency	hertz	Hz	s^{-1}	
Force	newton	N	$kg \cdot m/s^2$	J/m
Pressure	pascal	Pa	$kg/m \cdot s^2$	N/m^2
Energy, Work	joule	J	$kg \cdot m^2/s^2$	$N \cdot m$
Power	watt	W	$kg \cdot m^2/s^3$	J/s
Electric charge	coulomb	C	$A \cdot s$	
Electric potential	volt	V	$kg \cdot m^2/A \cdot s^3$	W/A
Capacitance	farad	F	$A^2 \cdot s^4/kg \cdot m^2$	C/V
Electric resistance	ohm	Ω	$kg \cdot m^2/A^2 \cdot s^3$	V/A
Magnetic flux	weber	Wb	$kg \cdot m^2/A \cdot s^2$	$V \cdot s$
Magnetic field	tesla	T	$kg/A \cdot s^2$	
Inductance	henry	H	$kg \cdot m^2/A^2 \cdot s^2$	$T \cdot m^2/A$

Prefixes used in SI			
Prefixe	**Symbol**	**Factor**	**Number**
Yocto	y	10^{-24}	0.000000000000000000000001
Zepto	z	10^{-21}	0.000000000000000000001
Atto	a	10^{-18}	0.000000000000000001
Femto	f	10^{-15}	0.000000000000001
Pico	p	10^{-12}	0.000000000001
Nano	n	10^{-9}	0.000000001
Micro	μ	10^{-6}	0.000001
Mili	m	10^{-3}	0.001
Centi	c	10^{-2}	0.01
Deci	d	10^{-1}	0.1
Deca	da	10^{1}	10
Hecto	h	10^{2}	100
Kilo	k	10^{3}	1 000
Mega	M	10^{6}	1 000 000
Giga	G	10^{9}	1 000 000 000
Tera	T	10^{12}	1 000 000 000 000
Peta	P	10^{15}	1 000 000 000 000 000
Exa	E	10^{18}	1 000 000 000 000 000 000
Zetta	Z	10^{21}	1 000 000 000 000 000 000 000
Yotta	Y	10^{24}	1 000 000 000 000 000 000 000 000

Note: A mass of 1000 kg is also known as a ton (t).

Constants of the new SI			
Constant	**Symbol**	**Numerical value**	**Unit**
Hyperfine transition frequency of Cs	Δf_{Cs}	9 192 631 770	Hz
Velocity of light in vacuum	c	299 792 458	ms^{-1}
Planck constant	h	$6.626\ 070\ 15 \times 10^{-34}$	J s
Elementary charge	e	$1.602\ 176\ 634 \times 10^{-19}$	C
Boltzmann constant	k	$1.380\ 649 \times 10^{-23}$	$\mathrm{J\ K}^{-1}$
Avogadro constant	N_{A}	$6.022\ 140\ 76 \times 10^{23}$	mol^{-1}
Luminous efficacy	K_{cd}	683	$\mathrm{lm\ W}^{-1}$

Note: $\Delta f_{\mathrm{Cs}} = \Delta \nu_{\mathrm{Cs}}$ is the unperturbed ground state hyperfine transition frequency of the Caesium 133 isotope.

In the next figure, one can see the relation between the new units of the International System and the fixed constants.

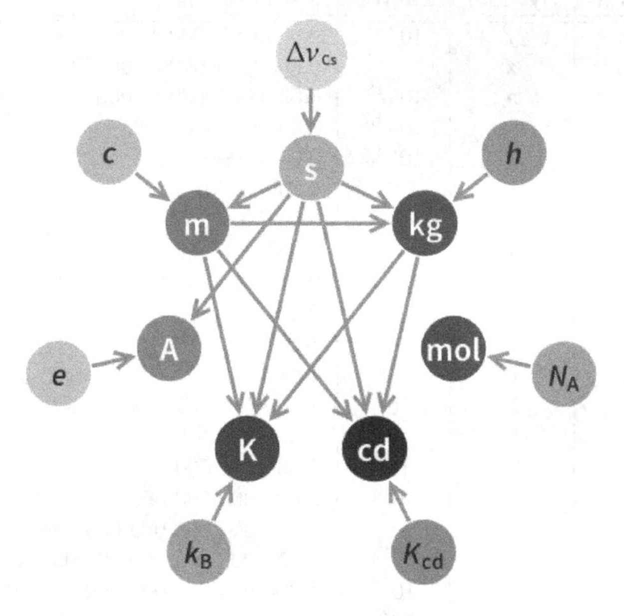

B.1 Homework

Estimate the Planck constant h with a simple voltmeter!

Indication: Use LEDs of different colors, consider the (inner) photoelectric effect

$$E = hf = e\varphi,$$

and delineate (Kotabage, 2019) the threshold voltage φ for various frequencies f or wave lengths $\lambda = c/f$.

References

BIPM (2019). "A concise summary of the International System of Units, SI", *SI Brochure*, 9th edition.

Kotabage, C. (2019). "Can a glowing LED light the road to accurate determination of Planck's constant?", *Resonance* **24**, 767–770.

Bibliography[1]

Allan, D.W. *et al.* (1997). *The Science of Timekeeping* (Hewlett Packard Applicate Note 1289).

Arnold, D.N. and Rogness, J. (2008). "Möbius transformations revealed", *Notices AMS* **55**(10), 1226.

Ashby, N. (2002). "Relativity and the global positioning system", *Phys. Today* **55**(5), 41–51; *Living Rev. Relativ.* **6** (2003) 1.

Autschbach, J. (2014). "Relativistic calculations of magnetic resonance", *Phil. Trans. R. Soc. A* **372**, 20120, 489.

Bailey, K. *et al.* (1977). "Measurements of relativistic time dilatation for positive and negative muons in a circular orbit", *Nature* **268**, 301–305.

Barnett, S.M. (2010). "Resolution of the Abraham–Minkowski dilemma", *Phys. Rev. Lett.* **104**, 070401.

Batelaan, H. and Tonomura, A. (2009). "The Aharonov–Bohm effects: Variations on a subtle theme", *Phys. Today* **62**(9), 38.

Béché, A. *et al.* (2014). "Magnetic monopole field exposed by electrons", *Nat. Phys.* **10**, 26–29.

Becker, W., Kramer, M. and Sesana, A. (2018). "Pulsar timing and its application for navigation and gravitational wave detection", *Space Sci. Rev.* **214**, 30.

Bender, P.L. *et al.* (1973). "The Lunar laser ranging experiment", *Science* **182**, 229–238.

Berkovits, N. *et al.* (2017). *Memorial Volume for Stanley Mandelstam* (World Scientific, Singapore).

Bethe, H.A. (2000). "The German uranium project", *Phys. Today* **July**, 34.

Bleyer, U. *et al.* (1979). "Zur Geschichte der Lichtausbreitung", *Die Sterne* **55**, 24–40. [On the history of light propagation, in German]

Bell, J.S. and Weaire, D. (1992). "George Francis FitzGerald", *Phys. World* **September**, 31–35.

[1]This list includes suggestions for a further reading.

BIPM (2019). "SI. A concise summery of the international system of units", *SI Brochure*, 9th edition.

Bobis, L. and Lequeux, J. (2008). "Cassini, Rømer and the velocity of light", *J. Astron. History Heritage* **11**(2), 97–105.

Bocquet, J.-P. *et al.* (2010). "Limits on light-speed anisotropies from Compton scattering of high-energy electrons", *Phys. Rev. Lett.* **104**, 241601.

Boran, S. *et al.* (2018). "GW170817 falsifies dark matter emulators", *Phys. Rev. D* **97**, 041501.

Brewer J.-S. *et al.* (2019). "$^{27}Al^{+}$ Quantum-logic clock with a systematic uncertainty below 10^{-18}", *Phys. Rev. Lett.* **123**, 03320.

Casado de Lucas, D. (2008). *Pequeña Biografía de Albert Einstein* [Short biography of Albert Einstein, in Spanish].

Cercignani, C. and Kremer, G.M. (2002). *The Relativistic Boltzmann Equation: Theory and Application* (Birkhäuser Verlag, Basel).

Chou, C.W. *et al.* (2010). "Optical clocks and relativity", *Science* **329**, 1630–1633.

Cigan, P. *et al.* (2019). "High angular resolution ALMA images of dust and molecules in the SN 1987 A ejecta", *Astrophys. J.* **886**, 51.

Ciufolini, I. *et al.* (2012). "Overview of the LARES mission: Orbit, error analysis and technological aspects", *Phys.: Conf. Ser.* **354**, 012002; *Eur. Phys. J.* **76**, 120 (2016).

Dehghani, M. (2009). "Gupta-Bleuler quantization for massive and massless free vector fields", *Braz. J. Phys.* **39**(3), 559.

Deryabin, M.V. and Pustyl'nikov, L.D. (2003). "Generalized relativistic billiards", *Regular Chaotic Dynam.* **8**(3).

D' Inverno, R. (1995). *Introducing Einstein's Relativity* (Claredon Press, Oxford).

Doat, T., Parizot, E. and Vézien, J. "A carom billiard to understand special relativity", in: *2011 IEEE Virtual Reality Conference*, Singapore, 2011, pp. 201–202, doi: 10.1109/VR.2011.5759468.

Droin, C. *et al.* (2019). "Low-dimensional dynamics of two coupled biological oscillators", *Nat. Phys.* **15**, 1086.

Dunkel, J. *et al.* (2009). "Nonlocal observables and lightcone-averaging in relativistic thermodynamics", *Nat. Phys.* **5**, 741–747.

Einstein, A. (1905a). "Uber einen die Erzeugung und Verwandlung des Lichtes betreffenden heuristischen Gesichtspunkt", *Annalen der Physik* **17**, 132–148 [On a Heuristic Point of View about the Creation and Conversion of Light, in German].

Einstein, A. (1905b). "Zur Elektrodynamik bewegter Körper", *Annalen der Physik* **17**(10), 891–921 [On the Electrodynamics of Moving Bodies, in German].

Einstein, A. (1916). *Relativity — The Special and General Theory* (Princeton University Press) (100 Anniversity Edition, 2019).

Essén, H. (2002). "Note on the relativistic elastic head-on collision", *Eur. J. Phys.* **23**, 565–568.

Fearn, H. (2007). "Can light signals travel faster than c in nontrivial vacua in flat space-time? Relativistic causality. II", *Laser Phys.* **17**, 695.

Feynman, R.P. (1988). *QED — The Strange Theory of Light and Matter* (Princeton University Press).

Franz, W. (1939). "Elektroneninterferenzen im Magnetfeld", *Verhandlungen der Deutchen Physikalischen Gesellschaft* **65**; "Über zwei unorthodoxe Interferenzversuche", *Zeitschrift für Physik* **184**, 85–91 (1965) [Electron interference in the magnetic field; On two unorthodox probes of interference, in German].

Freund, F. (2008). *Special Relativity for Beginners: A Textbook for Undergraduates* (World Scientific, Singapore).

Frisch, D.H. and Smith, J.H. (1963). "Measurement of the relativistic time dilation using μ-mesons", *Am. J. Phys.* **31**(5), 342–355.

Gabrielse, G. (2013). "The standard models's greatest triumph", *Phys. Today* **December**, 64–65.

Giulini, D. (2001). "Uniqueness of simultaneity", *Brit. J. Philos. Sci.* **52**, 651–670.

Goldhaber, A.S. and Nieto, M.M. (2010). "Photon and graviton mass limits", *Rev. Mod. Phys.* **82**, 939–979.

Goldstein, H. (1980). *Classical Mechanics*, 2nd ed. (Addison-Wesley).

Gourgoulhon, E. (2013). *Special Relativity in General Frames*, Graduate texts in Physics (Springer, Berlin).

Grant, A. (2019). "What it took to capture a black hole", *Phys. Today Res. Technol.* 11 April 2019.

Haas, R. (2021). "Clock comparison using black holes", *Nat. Phys.* **17**, 164.

Hafele, J.C. and Keating, R.E. (1972). "Around-the-world atomic clocks: Predicted relativistic time gains; observed relativistic time gains", *Science* **177**(4044) 166–168; 168–170.

Hartnett, J.G. and Luiten, A. (2011). "Colloquium: Comparison of astrophysical and terrestrial frequency standards", *Rev. Mod. Phys.* **83**, 1–8.

Hehl, F.W. *et al.* (1991). "Two lectures on fermions and gravity", in: *Geometry and Theoretical Physics*, Bad Honnef Lectures, 12–16 February 1990. J. Debrus and A.C. Hirshfeld eds. (Springer, Berlin, 1991), pp. 56–140.

Hermann, A. (1997). *Einstein. En privado* (Ediciones Temas de Hoy, Madrid) 583 pp. [Einstein privately. in Spanish]

Herrmann, S. *et al.* (2009). "Rotating optical cavity experiment testing Lorentz invariance at the 10^{-17} level", *Phys. Rev. D* **80**, 105011.

Illy, J. (2012). *The Practical Einstein: Experiments, Patents, Inventions* (Johns Hopkins University Press, Baltimore) (Book review: *The Brit. J. History Sci.* (2014) **47**, 382–383.)

Itzkinson, C. and Zuber, J.-B. (1980). *Quantum Field Theory* (MacGraw–Hill, New York).

Ives, H.E. and Stilwell, G.R. (1938). "An experimental study of the rate of a moving clock", *J. Opt. Soc. Am.* **28**, 215–226; *J. Opt. Soc. Am.* **31**, 369–374.

Iwanenko, D. and Pomeranchuk, I. (1944). "On the maximal energy attainable in a Betatron", *Phys. Rev.* **65**, 343.

Jester, S. (2008). "Retardation, magnification and the appearance of relativistic jets, *Month. Not. Royal Astron. Soc.* **389**, 1507.

Kibble, T.W.B. (1960). "Kinematics of general scattering processes and the Mandelstam representation", *Phys. Rev.* **117**, 1159.

Kiehn, R.M. (2001). "Falaco solitons, cosmic strings in a swimming pool", arXiv: gr-qc/0101098.

Kittel, C. (1974). "Larmor and the prehistory of the Lorentz transformations", *Am. J. Phys.* **42**, 726–729.

Kotabage, C. (2019). "Can a glowing LED light the road to accurate determination of Planck's constant?", *Resonance* **24**, 767–770.

Kraus, U. (2000). "Brightness and color of rapidly moving objects: The visual appearance of large sphere revisited", *Am. J. Phys.* **68**, 56–60; "First-person visualizations of the special and general theory of relativity", *Eur. J. Phys.* **29** (2008) 1–13.

Lahaye, T. *et al.* (2012). "Fizeau's "aether-drag" experiment in the undergraduate laboratory", *Am. J. Phys.* **80**(6), 497.

Lampa, A. (1924). "Wie erscheint nach der Relativitätstheorie ein bewegter Stab einem ruhenden Beobachter?", *Z. Physik* **27**, 138–148 [According to the theory of relativity, how does a moving rod appear to a non-moving observer?, in German].

Larmor, J. (1897). "On a dynamical theory of the electric and luminiferous medium, III", *Phil. Trans. Roy. Soc.* **190**, 205–300.

Lefévre, W. (2007): *Inside the Camera Obscura: Optics and Art Under the Spell of the Projected Image* (Max-Planck Institute for the History of Science, Berlin).

Lemke, J., Mielke, E.W. and Hehl, F.W. (1994). "Äquivalenzprinzip für Materiewellen? — Experimente mit Neutronen, Atomen, Neutrinos", *Physik in unserer Zeit* **25**, 36–43 [Equivalence principle for matter waves? — experiments with neutrons, atoms, neutrinos, in German].

Lewis, G.N. and Tolman, R.C. (1909). "The principle of relativity, and non-Newtonian mechanics", *Proc. Am. Acad. Arts Sci.* **44**, 709–726.

Liebscher, D.E. (1977). *Relativitätstheorie mit Zirkel und Lineal* (Friedr. Vieweg & Sohn, Braunschweig) [Theory of relativity with circle and ruler, in German.]

Lombardi, M.A. *et al.* (2007). "NIST primary frequency standards and the realization of the SI second", *The J. Measure. Sci.* **2**(4).

Lyons, L. (2012). "Discovery or fluke: Statistics in particle physics", *Phys. Today* **July**, 45–51.

Mac Cullagh, J. (1846). "An essay towards a dynamical theory of cristaline reflexion and refraction" (read December 9th, 1839); *Dublin Trans. Roy. Irish Acad.* (Dublin) **21**, 17–50.

Manko, V.S. *et al.* (2000). "Exact solution for the exterior field of a rotating neutron star", *Phys. Rev.* **61**, 081501 (Rapid Communication).

Martinez, A. (2005). "Handling evidence in history: The case of Einstein's wife", *School Sci. Rev.* **86**(316), 49–56.

Matt Leone, R. (2002). *Billiards in Space: The Study of Classical Relativistic Collisions*, PHY600 Special Relativity Theory (University of Arizona, Flagstaff).

Mayos, G. *et al.* (2008) *D'Alembert. Vida, Obra y Pensamiento* (Planeta DeAgostini, Barcelona) [D'Alembert. Life, work and thought, in Spanish].

Mielke, E.W. (1977). "Knot wormholes in geometrodynamics?", *Gen. Rel. Grav.* **8**, 175–196 [reprinted in: *Knots and Applications*, L.H. Kauffman, ed. (World Scientific, Singapore 1995), pp. 229–250].

Mielke, E.W. (1980). "Magnetische Monopole in vereinheitlichten Eichtheorien", *Z. Naturforsch.* **41a**, 777–787 [Magnetic monopoles in unified gauge theories, in German].

Mielke, E.W. (1997). *Sonne, Mond und ... Schwarze Löcher* (Friedr. Vieweg & Sohn, Braunschweig/Wiesbaden [Springer Book Archives]), 282 pages and 12 color plates.[Sun, Moon and ... Black Holes, in German].

Mielke, E.W. and Hehl, F.W. (1988). *Die Entwicklung der Eichtheorien: Marginalien zu deren Wissenchaftsgeschichte*, in: *Exakte Wissenschaften und ihre philosophische Grundlegung — Vorträge des Internationalen Hermann-Weyl-Kongresses*, Kiel 1985, W. Deppert, K. Hübner, A. Oberschelp und V. Weidemann (Hrsg.), (Verlag Peter Lang, Frankfurt a. M.), pp. 191–231. [The Development of Gauge Theories: Marginals to Their History of Science, in German].

Mielke, E.W. and Miguel A. Marquina Carmona (2013). "Relativity and the tunneling problem in a "reduced" waveguide", *Int. J. Optics* **2013**, Article ID 947068 [Hindawi], 10 pages.

Minkowski, H. (1908). *Raum und Zeit* (Space and Time, Address delivered at the 80th Assembly of German Natural Scientists and Physicians, Cologne, 21 September 1908). *Physikalische Zeitschrift* **9**, 104–111.

Mirabel, I.F. and Rodríguez, L.F. (1998). "Microquasars in our galaxy", *Nature* **392**, 673.

Misner, C.W., Thorne, K.S. and Zurek, W.H. (2009). "John Wheeler, relativity, and quantum information", *Phys. Today* **April**, 40–46.

Moser, M.-B. and Moser, E.I. (2016). "The brain's GPS tells you where you are and where you've come from", *Sci. Am.* **314**.

Müller, T. *et al.* (2011). "Special relativistic visualization by local ray tracing", *IEEE Trans. Visualiz. Comput. Graph.* **January**.

Müller, T. and Weiskopf, D. (2011). "Special-relativistic visualization", *Comput. Sci. Eng.* **July/August**, 85–93.

Naumann, R. and Stroke, H. (1996). "Einstein and the atomic clock", *Phys. World* **April**, 76.

Nimtz, G. (2011). "Tunneling confronts special relativity", *Found. Phys.* **41**, 1193–1199.

Okun, L.B. (1989). "The concept of mass", *Phys. Today* **31**, 31–36.

Oliva, G.A. and F. Frutos-Alfaro (2021). "Effects of the treatment of the mass quadrupole moment on ray-tracing applications for rapidly rotating neutron stars", *Monthly Notices of the Royal Astronomical Society*, **505** Issue 2, August 2021, pp. 2870–2885.

Pais, A. (1982). *Subtle is the Lord: The Science and Life of Albert Einstein* (Oxford University Press).

Parsons, P. (2011). *3 — Minute Einstein* (Art Blume, Barcelona).

Pascoli, G. (2017). "The Sagnac effect and its interpretation by Paul Langevin", *Comptes Rendus Physique* **18**, 563–569.

Patt, H.J. and Nemec, P. (2000). *Relativity for Windows* (Springer-Verlag, Berlin-Heidelberg 2000).

Penrose, R. (1959). "The apparent shape of a relativistically moving sphere", *Proc. Cambridge Phil. Soc.* **55**, 137.

Poincaré, H. (1904). *The Principles of Mathematical Physics*, The Foundations of Science (Science Press, New York 1906) pp. 297–320.

Pospelov, M. and Romalis, M. (2004). "Lorentz invariance on trial", *Phys. Today* **July**, 40–46.

Pound, R.V. and Rebka, G.A. (1960). "Apparent weight of photons", *Phys. Rev. Lett.* **4**, 337–341.

Pound, R.V. and Snider, J.L. (1965). "Effect of gravity on gamma radiation", *Phys. Rev.* **140**, B788–B803.

Riemann, B. (1867). "Ein Beitrag zur Elektrodynamik", *Annalen der Physik und Chemie* **131**, 237–243 [A contribution to electrodynamics, in German].

Rindler, W. (1961). "Length contraction paradoxes", *Amer. J. Phys.* **29**, 365.

Rindler, W. (1991). *Introduction to Special Relativity*, 2nd ed. (Clarendon Press, Oxford).

Rossi, B. and Hall, D.B. (1941). "Variation of the rate of decay of mesotrons with momentum", *Phys. Rev.* **59**(3), 223–228.

Rothman, T. (2006). "Lost in Einstein's shadow", *Am. Sci.* **94**, 112–113.

Ruder, H. and M. (1993). *Die Spezielle Relativitätstheorie* (Friedr. Vieweg & Sohn, Braunschweig/Wiesbaden) [The Special Theory of Relativity, in German].

Schwartz, H.M. (1970). "Generalization of an elementary formula in relativistic kinematics due to Pauli", *Am. J. Phys.* **38**, 927–929.

Serway, R.A. y J.W. Jewett, Jr. (2008). *Physics for Scientist and Engineers with Modern Physics*, Vol. II. 9th ed. (Brooks/Cole).

Sexl, R. and Schmidt, H.K. (1979). *Raum-Zeit-Relativität* (Friedr. Vieweg & Sohn, Braunschweig) [Spacetime Relativity, in German].

Sexl, R. and Urbandtke, H. (2001). *Relativity, Groups, Particles*, 4th ed. (Springer Wien, New York).

Sfarti, A. (2010). "Improved tests of special relativity via light speed anisotropy measurement", *Mod. Phys. Lett. A* **25**, 125.

Shankland, R.S. (1974). "Michelson and his interferometer", *Phys. Today*, **April**, 37–43.

Silberstein, L. (1912). "Quaternionic form of relativity", *Phil. Mag.* **23**, 790–809.

Smith, G.S. (2011). "Visualizing special relativity: The field of an electric dipole moving at relativistic speed", *Eur. J. Phys.* **32**, 695–710.

Snios, B. *et al.* (2019). "Detection of superluminal motion in the X-ray jet of M87", *Astrophys. J.* **879**(8).

STAR Collaboration (2011). "Observation of the antimatter helium-4 nucleus", *Nature* **473**, 353–356.

Stachel, J. (1996). "Albert Einstein and Mileva Maric: A collaboration that failed to develop", in: *Creative Couples in the Sciences*, H.M. Pycior, N.G. Slack, and P.G. Abir-Am, editors (Rutgers University Press).

Stentz, A. *et al.* (2002). "A system for semi-autonomous tractor operations", *Auton. Robots* **13**, 87.

Stewart, A.B. (1964). "The discovery of stellar aberration", *Sci. Am.* **210**, 100–108.

Strain, R.M. (2010). "Asymptotic stability of the relativistic Boltzmann equation for the soft potentials", *Commun. Math. Phys.* **300**, 529–597.

Straumann, N. (1990). *Spezielle Relativitätstheorie* (Skriptum, University of Zurich) [Special Relativity Theory, in German].

Taylor, E.F. and Wheeler, J.A. (1992). *Spacetime Physics: Introduction to Special Relativity*, 2nd ed. (W. H. Freeman and Company, New York).

Taylor S.R. *et al.* (2016). "Are we there yet? Time to detection of nanohertz gravitational waves based on pulsar-timing array limits", *Astrophys. J. Lett.* **819**, L6.

Terrell, J. (1959). "Invisibility of the Lorentz contraction", *Phys. Rev.* **116**, 1041; "The Terrell effect", *Am. J. Phys.* **57**(1), 9.

Tixaire, A.G. (2006). "Relatividad, Tiempo y Asuntos de Gravedad", *Rev. R. Acad. Cienc. Exact. Fís. Nat.* (Esp.) **100**(1), 141–155 [Relativity, Time and issues of gravity, in Spanish].

Tonomura, A. (2005). "Direct observation of thitherto unobservable quantum phenomena by using electrons", *PNAS* **102**(42), 14952–14959.

Van Baak, T. (2007). "An adventure in relative time-keeping", *Phys. Today Lett.* **60**, 16.

Van den Brand, J. (2019). "A primer on LIGO and Virgo gravitational wave detection", *European School of High-Energy Physics*, St. Petersburg, September 16, 2019.

Van Oss, Rosine G. (1983). "D'Alembert and the fourth dimension", *Historia Mathematica* **10**, 455–457.

Vessot, R.F.C. *et al.* (1980). "Test of relativistic gravitation with a space-borne hydrogen maser", *Phys. Rev. Lett.* **45**, 2081.

Vincent, F.H. *et al.* (2011). "GYOTO: A new general relativistic ray-tracing code", *Class. Quantum Grav.* **28**, 225011.

Vollmer, M. (2004). "Physics of the microwave oven", *Phys. Education* **39**, 74–81.

Voigt, W. (1887). *Theorie des Lichtes für bewegte Medien*, Nachrichten von der Königl. Gesellschaft der Wissenschaften und der Georg-August-Universität zu Göttingen, No. 8, pp. 177–238 [Theory of light for moving media, in German].

Von Laue, M. (1907). "Die Mitführung des Lichtes durch bewegte Körper nach dem Relativitätsprinzip", *Ann. Phys.* **328**, 989–990 [The Dragging of Light by Moving Bodies in Accordance with the Principle of Relativity, in German].

Von Laue, M. (1913). *Das Relativitätsprinzip*, 2nd ed. (Friedr. Vieweg & Sohn, Braunschweig) [The Relativity principle, in German].

Von Laue, M. (1948). Dr. Ludwig Lange (1863–1936). "Ein zu Unrecht Vergessener", *Naturwissenschaften* **35**(7), 193–196 [An unjustly forgotten, in German].

Weiskopf, D. (2010). "A survey of visualization methods for special relativity", in: *Scientific Visualization: Advanced Concepts*, Hans Hagen, editor (Dagstuhl Publishing, Schloss Dagstuhl — Leibniz Center for Informatics, Germany), pp. 289–302.

Weisskopf, V. (1960). "The visual appearance of rapidly moving objects", *Phys. Today* **13**, 24 (September 1960).

Wehrle, A.E. *et al.* (2009). *What is the Structure of Relativistic Jets in AGN on Scales of Light Days?*, Whitepaper.

Wheeler, J.A. (1968). "Einstein's vision", *Phys. Today* **23**, May (1970), 68.

Will, C.M. (2006). "Special relativity: A centenary perspective", in: *Proceedings: Einstein, 1905–2005: Poincaré Seminar 2005*, T. Damour, O. Darrigol, B. Duplantier and V. Rivasseau, editors (Birkhäuser, Basel), pp. 33–58.

Wille, K. (2013). "Synchrotron radiation", JUAS (Joint Universities Accelerator School) 28 January–1 February.

Williams, J.G. *et al.* (2012). "Lunar laser ranging tests of the equivalence principle", *Class. Quant. Grav.* **29**, 184004.

Wipf, A. (2007). *Elektrodynamik* (Skriptum, University of Jena) [Electrodynamics, in German].

Wynands, R. and Weyers, S. (2005). "Atomic fountain clocks", *Metrologia* **42**(3).

Yang, Y.Y. *et al.* (2019). "Gravitational wave GW170817: A new-born sub-millisecond pulsar and the properties of coalescing double neutron stars", *New Astron.* **70**, 51–56.

Index

CPSIA information can be obtained
at www.ICGtesting.com
Printed in the USA
JSHW011804090322
23577JS00002B/7